# HOME REPAIR AND IMPROVEMENT

# HEATING AND COOLING

**TIME® LIFE BOOKS**

*Other Publications*
THE TIME-LIFE COMPLETE GARDENER
JOURNEY THROUGH THE MIND AND BODY
WEIGHT WATCHERS® SMART CHOICE RECIPE COLLECTION
TRUE CRIME
THE AMERICAN INDIANS
THE ART OF WOODWORKING
LOST CIVILIZATIONS
ECHOES OF GLORY
THE NEW FACE OF WAR
HOW THINGS WORK
WINGS OF WAR
CREATIVE EVERYDAY COOKING
COLLECTOR'S LIBRARY OF THE UNKNOWN
CLASSICS OF WORLD WAR II
TIME-LIFE LIBRARY OF CURIOUS AND UNUSUAL FACTS
AMERICAN COUNTRY
VOYAGE THROUGH THE UNIVERSE
THE THIRD REICH
MYSTERIES OF THE UNKNOWN
TIME FRAME
FIX IT YOURSELF
FITNESS, HEALTH AND NUTRITION
SUCCESSFUL PARENTING
HEALTHY HOME COOKING
UNDERSTANDING COMPUTERS
LIBRARY OF NATIONS
THE ENCHANTED WORLD
THE KODAK LIBRARY OF CREATIVE PHOTOGRAPHY
GREAT MEALS IN MINUTES
THE CIVIL WAR
PLANET EARTH
COLLECTOR'S LIBRARY OF THE CIVIL WAR
THE EPIC OF FLIGHT
THE GOOD COOK
WORLD WAR II
THE OLD WEST

*For information on and a full description*
*of any of the Time-Life Books series listed above,*
*please call 1-800-621-7026 or write:*
Reader Information
Time-Life Customer Service
P.O. Box C-32068
Richmond, Virginia 23261-2068

# HEATING AND COOLING

BY THE EDITORS OF TIME-LIFE BOOKS, ALEXANDRIA, VIRGINIA

*The Consultants*
Greg M. Cumming holds Virginia Class A licenses in electrical, mechanical, and plumbing contracting and in gas fitting. He graduated with high honors from Northern Virginia Community College in 1986, with a major in heating and air conditioning. He owns and operates Accurate Air, Inc., in Fairfax.

Jeff Palumbo is a registered journeyman carpenter who has a home-building and remodeling business in northern Virginia. His interest in carpentry was sparked by his grandfather, a master carpenter with more than 50 years' experience. Palumbo teaches in the Fairfax County Adult Education Program.

Mark M. Steele is a professional home inspector in the Washington, D.C., area. He has developed and conducted training programs in home-ownership skills for first-time homeowners. He appears frequently on television and radio as an expert in home repair and consumer topics.

# CONTENTS

# A Climate Made to Order

The goal of any heating or cooling system is to provide a comfortable environment in every room of the house. The most common defect—too much heat in one place and not enough in another—may be corrected merely by adding a damper to a duct or replacing a valve in a hot-water convector. Effective control of the system is also essential; a properly functioning thermostat will keep the temperature where you want it.

Cleaning the contacts on a thermostat →

# Balancing Your Forced-Air System

**A** heating system should provide each room in the house with just the right amount of warmth. This does not necessarily mean uniformity: You may want more heat delivered to bathrooms or a nursery, and less to bedrooms and the kitchen. Adjusting a forced-air system is easy if there are dampers—the movable metal plates in ducts that lead from the furnace to the room registers.

**Adjusting the Dampers:** Depending on the duct pattern, dampers may be clustered close to the furnace or scattered widely, as in the example shown below. Whatever the arrangement, all dampers work alike: When the damper handle parallels the duct path, the duct is wide open; when the handle is perpendicular, the duct is closed; and when the handle is in between, the duct is partially closed, or damped.

Balancing is simple, but it may take several days. Start by damping the duct to a room that seems too hot—preferably one that lies close to the furnace. Wait 6 to 8 hours, then check the temperature of the room, by "feel" and with a thermometer held 4 or 5 feet above the floor. Repeat this process, one room at a time, until you have worked your way through the house.

You may find that the room containing the thermostat heats up so quickly that the furnace shuts down before other rooms are fully heated. The solution is to damp the duct leading to the thermostat room—almost completely, if necessary. More difficult is the situation in which the rooms farthest from the furnace never reach a comfortable temperature even with their ducts wide open. Try stepping up the speed of the furnace fan *(page 36)* to send more air to these distant rooms.

**Fine-Tuning Your System:** When balancing is complete, you will almost certainly have to make minor adjustments, since increasing or decreasing the amount of air flowing into one room affects all the others. Go through the rooms again, adjusting one damper at a time and observing the 6-to-8-hour waiting period. Finally, when every room is getting the right amount of heat, mark the damper settings with a felt-tip pen. (If you have central air conditioning, repeat the entire operation in the summer and label the different settings HEAT and A/C.)

**Installing Your Own Dampers:** If your heating system lacks dampers, you can add them yourself. For round duct, buy a matching 2-foot-long section with a factory-installed damper. With the furnace off, insert the new section at the beginning of a duct run, using the method explained on pages 60 and 61.

Ready-made damper sections for rectangular ducts are harder to find because the ducts come in many sizes. The steps at right show you how to make and install your own.

---

**TOOLS**

| | |
|---|---|
| Tin snips | Screwdriver |
| Broad-billed pliers | Pliers |
| Ball-peen hammer | |

**MATERIALS**

Sheet metal
Spring-loaded clips
Duct tape

**SAFETY TIPS**

*When you are cutting and handling sheet metal, wear heavy work gloves to protect yourself against sharp edges.*

### Anatomy of a forced-air duct system.

Duct runs begin at the plenum, a large chamber attached to the furnace. An extended-plenum system *(below)* has a main duct (or ducts) running from the plenum, and branch ducts running from a main to the registers in each room. You can usually detect the route and destination of a duct run by visual inspection; if necessary, close a damper to see which room turns cold. Dampers are found near the starts of the branches.

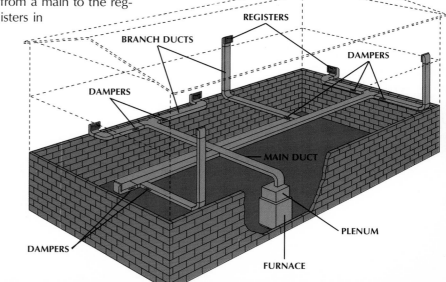

# A DAMPER FOR A RECTANGULAR DUCT

## 1. Making the damper.

◆ Measure the height and width of the duct.

◆ Wearing work gloves, use tin snips to cut a rectangle of sheet metal 1 inch longer and wider than these dimensions.

◆ Mark the dimensions of the duct within this rectangle, snip off the corners at about a 45-degree angle, and with broad-billed pliers fold the edges of the metal back along the marked lines to form rounded edges that are two layers thick.

◆ Hardware for the damper is available at heating-supply and hardware stores; buy the type with two spring-loaded clips that do not require welding or riveting.

◆ Slide the clips over the short edges of the damper, lining them up at the exact center. Set the damper on a firm surface and drive the clip prongs through the damper with a ball-peen hammer.

## 2. Opening the duct.

◆ Turn the furnace off and let it cool for an hour or so.

◆ Remove any duct tape over a connection of the branch duct, just past the point where the branch leaves the main duct, and open this connection.

◆ If the duct does not simply snap apart, remove the horizontal S clips and vertical drive clips. With a screwdriver, open the tabs at the tops and bottoms of the drive clips, then pull the clips down and off the duct connection with pliers. Separate the duct sections by pulling them out of the S clips.

◆ Remove the hanger supporting the duct section that lies farther from the main duct, and carefully lower the free end of this section until it is clear; support it in this position on a convenient prop—the rung of a ladder, for example.

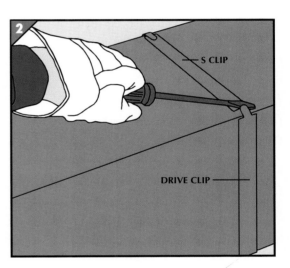

S CLIP

DRIVE CLIP

## 3. Installing the damper.

◆ Mark dots on the sides of the lowered duct section at a distance from its end that is equal to half the height of the damper plus 2 inches. Draw vertical lines through the dots. Draw horizontal lines along the centers of the sides of the duct.

◆ Where the lines intersect, drill holes the size of the bolts on the damper clips.

◆ Compress the spring-loaded clips, slide the damper into the duct, and release the bolts into the drilled holes. Install the damper handle.

◆ To rejoin the ducts, slip their edges into the S clip. Fold the bottom tabs of the drive clips, tap these clips lightly into place with a hammer, then fold the top tabs. Cover the connection with duct tape.

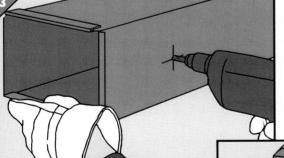

The strategy for tuning the temperatures of individual rooms heated by hot-water convectors is much the same as that for balancing a forced-air system *(pages 8-9)*. The major difference is that hot-water systems cannot be balanced as precisely.

**Directing the Flow:** If your house has more than one thermostat, the system probably balances itself automatically. Otherwise, your system will have valves: balancing valves in the basement that correspond to warm-air dampers, or inlet valves at the convectors that correspond to the movable vanes of registers.

**Tuning Up the System:** At the beginning of the heating season—and always before adjusting a manually balanced system—vacuum convector fins and straighten them if they are bent. Then bleed the air out of the convectors to ensure that each is completely full of water for efficient heating. During the season, bleed any convector that seems cooler than normal.

**TOOLS**
Screwdriver
Broad-billed pliers
Felt-tip pen

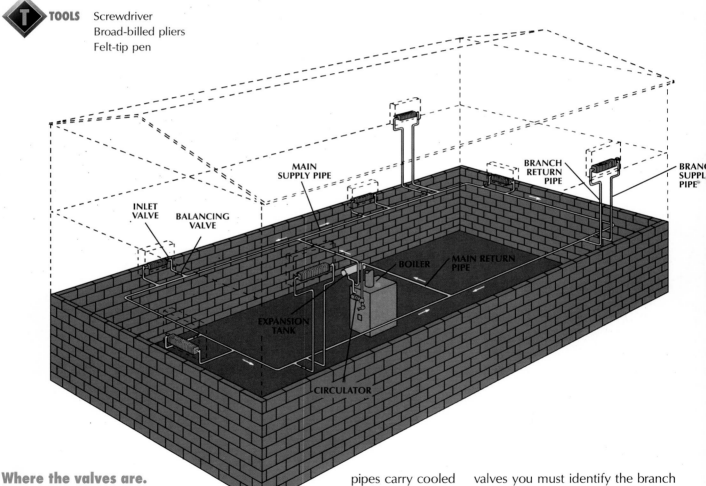

MAIN
SUPPLY PIPE

BRANCH
RETURN
PIPE

BRANC
SUPPL
PIPE

INLET
VALVE   BALANCING
VALVE

BOILER   MAIN RETURN
PIPE

EXPANSION
TANK

CIRCULATOR

## Where the valves are.

In a typical one-pipe hot-water system like the one above (variations of this basic design are shown on page 75), a circulator pump draws hot water from a boiler and distributes it through main and branch supply pipes to baseboard heaters (or, in some systems, to taller units called convectors). Return pipes carry cooled water back to the boiler to be reheated. If your system has balancing valves *(red),* you will find them wherever the main line branches to a convector; inlet valves *(blue)* are located at one end of each convector. Either can be used to balance the system, but if you have balancing valves you must identify the branch lines by room. To do so, close all the valves on a cool day and set the thermostat at 68°F. Open a valve and wait about an hour while the convector it serves warms up. Tag that valve with the name of the room, then open the others, one by one, until the valves are all tagged.

## Straightening convector fins.

The metal fins on convectors should provide straight, smooth paths for air rising through the fixture. If a fin is bent, twist the metal back into shape with a pair of broad-billed pliers.

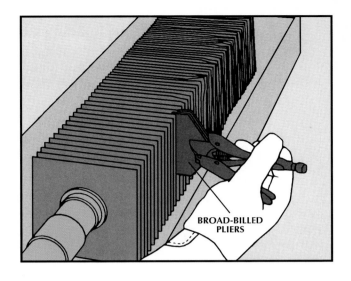

BROAD-BILLED PLIERS

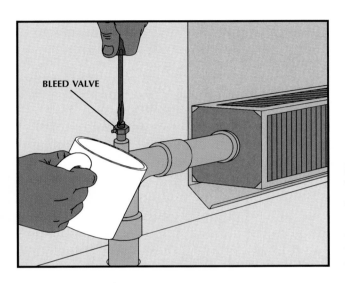

BLEED VALVE

## Bleeding a convector.

With the system running, hold a cup or absorbent rag under the bleed valve, and unscrew the valve slowly until you hear air hissing out. When all of the air is discharged, hot water will spurt out quickly; close the valve immediately.

## Quieting a Convector

Convector fins expand and contract with changes in temperature, causing them to rub noisily against the support brackets. To reduce this, place a piece of plastic—cut from an empty milk jug, for example—between the fins and the support bracket (below). The plastic will prevent the rubbing while allowing the fins to expand and contract.

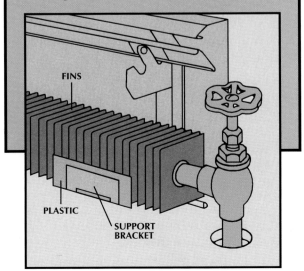

FINS

PLASTIC

SUPPORT BRACKET

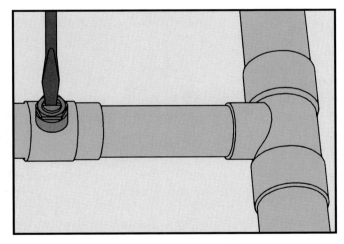

## Adjusting a balancing valve.

The screwdriver slot that controls this valve also indicates its setting: The valve is wide open when the slot is parallel to the path of the pipe (above), closed when the slot is perpendicular to the path. To regulate the flow of hot water to a convector, set the valve between these extremes, and when you have made all your balancing adjustments, mark the setting of the valve slot with a felt-tip pen.

Most home thermostats operate on low voltage and consist of three sections: a cover, a middle segment containing the temperature-sensing and -control mechanism, and a base plate or wall plate. Wire terminals and switches are typically located on the base plate or on the temperature-control section.

With time and use, all thermostat parts can malfunction. Often you need only adjust a jammed cover to get a balky furnace or air conditioner going again. Wires can work loose, and switches can acquire a coat of dust and lint.

**High-Voltage Thermostats:** Controls that are designed to regulate electric baseboard heaters operate on the same current as the heaters,

usually 220 volts. Known as line voltage thermostats, these models, which mount on a standard electric switch box, have just two parts: a removable cover and a control section with terminals. If your baseboard system has a wall switch, you can easily replace it with a line-voltage thermostat *(page 14)*. Otherwise, you will need to reroute the heater's power cable through a switch box before adding the thermostat.

**Upgrading:** Although simple repairs can keep a thermostat functioning for years, you may prefer to replace an older, manually set thermostat with a fuel-saving programmable model that can be set for different day- and nighttime tem-

peratures *(page 15)*. In most cases, this is a simple operation that involves little more than removing color-coded wires from the old thermostat and attaching them to color-coded terminals of the new one.

However, if you decide to change the location of a thermostat—to avoid drafts, for example—you will also need to run low-voltage cable containing the necessary number of 18-gauge wires to the new site.

Thermostats for furnaces and air conditioners usually have no more than five wires attached. A thermostat that has six or more wires is often connected to a heat pump, which requires a special thermostat. Contact your heat pump dealer for advice on a programmable replacement model.

**TOOLS**

| | |
|---|---|
| Screwdriver | Electric drill |
| Long-nose pliers | Hammer |
| Torpedo level | Wire strippers |

**MATERIALS**

| | |
|---|---|
| Alligator clips | 18-gauge wire |
| Nonsilicone switch cleaner | Insulation |
| | Wall anchors |
| Spackling compound | Wire caps |

## SIMPLE REPAIRS

### Tightening wire connections.
◆ Set the thermostat's main switch to the off position.
◆ Pull the cover gently from its support clips to reveal the temperature-control section. Tighten any terminal screws that you see.

◆ Remove the mounting screws that are holding the temperature-control section to the base plate, disclosing four or five additional terminal screws. Tighten them.

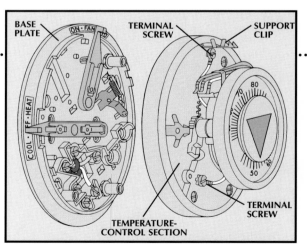

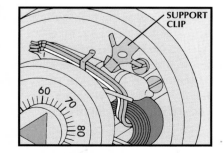

### Releasing jammed parts.
◆ When a thermostat dial binds, turn off the main switch. Remove the cover.
◆ Inspect the metal support clips that hold the cover in place. If one of the clips is bent *(left)*, use long-nose pliers to restore it to its original po-

sition or shape, as shown above.
◆ Replace the cover and set the switch for heating or cooling. Rotate the dial to be sure that it moves freely and turns the system on and off. If it does not, replace the thermostat *(page 15)*.

## Bypassing a Faulty Thermostat

Thermostats often fail at inconvenient times. To keep your house reasonably comfortable until you can replace the faulty control, bypass it with a jumper wire fitted with an alligator clip on each end. Remove the thermostat cover and the temperature-control section as shown on the preceding page. Next connect one end of the jumper to the terminal marked R or Rh and the other to terminal W. If the thermostat is at fault, the system will turn on. Turn it off when the temperature is comfortable by removing the jumper. Should the system not start with the jumper wire in place, look beyond the thermostat for the underlying difficulty.

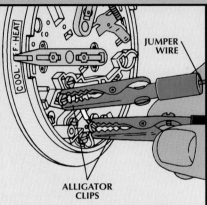

JUMPER WIRE

ALLIGATOR CLIPS

### Cleaning thermostat switches.

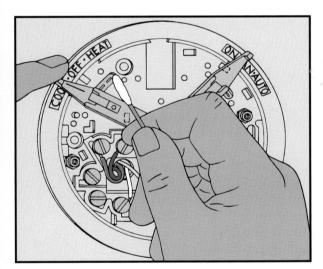

◆ Turn off power to the heating and cooling systems. Remove the thermostat cover and unscrew the temperature-control section from the base.
◆ Saturate a cotton swab with a nonsilicone switch cleaner, available at electronics stores, or a strong (50/50) vinegar-and-water solution, and clean the contacts near the switch levers. Move the levers from side to side to expose all the contacts.
◆ Remount the temperature-control section and attach the cover.

### Adjusting the anticipator.

This device, which contains an adjustable heating coil, keeps a thermostat from raising room temperature above the thermostat setting and keeps the furnace from turning on and off too often. To fix either problem, remove the thermostat cover and adjust the anticipator as follows (let the system adjust for a few hours after each change of setting):
◆ To correct swings of temperature more than 2°F above the thermostat setting, move the pointer of the anticipator down with the tip of a pencil, 0.1 ampere at a time.
◆ If the furnace starts and stops too frequently, move the pointer to a higher setting, 0.1 ampere at a time.

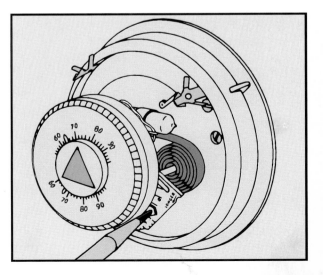

### Installing a line-voltage thermostat.

◆ At the service panel, shut off power to the baseboard heater.

◆ Remove the switch mounting screws and pull the existing switch from the box.

◆ Label the switch wires with the markings on the terminals: L1 and L2 for the wires to the service panel, T1 and T2 for wires to the heater. Disconnect the wires and set the switch aside.

◆ Locate the matching terminals on the back of the line-voltage thermostat and connect the wires as shown at right.

◆ Remove the cover, mount the thermostat on the box, and replace the cover.

**CAUTION** *To avoid damaging fragile thermostat components, leave the cover in place while connecting wires.*

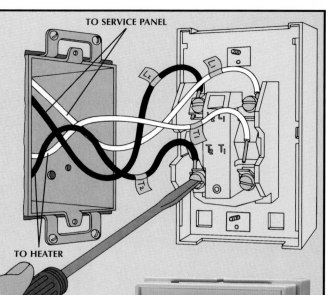

## A Cure for a Wayward Thermostat

Drafts can cause a thermostat to misread the temperature in a home's living spaces. A thermostat should be mounted on an interior wall where sunlight cannot strike it and away from direct air flow from windows, doors, and supply registers. However, even a properly located thermostat may still be exposed to drafts coming from inside the wall.

If your thermostat operates erratically, unscrew the base plate from the wall and fill unused mounting holes with spackling compound. Gently pack the hole around the low-voltage wiring with fiberglass or rock wool insulation *(below)*. Do not substitute caulk for insulation; you may need to pull the wires a little farther out of the wall later if you install a new thermostat.

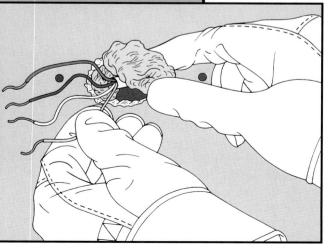

# INSTALLING A PROGRAMMABLE THERMOSTAT

## 1. Mounting the base.

◆ Unscrew the old thermostat from the wall and label each wire with the color code of the terminal to which it is attached. (These codes may not always match the color of the corresponding wire's insulation.)

◆ Position the new base plate on the wall, with the wires centered in the plate's wiring opening. Level the top of the plate with a torpedo level, and mark two mounting holes with a pencil (right).

◆ Set the base aside and drill $\frac{3}{16}$-inch holes at the marks. Tap wall anchors into the holes and screw the base to the wall.

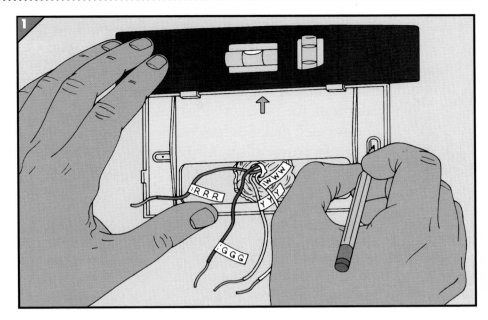

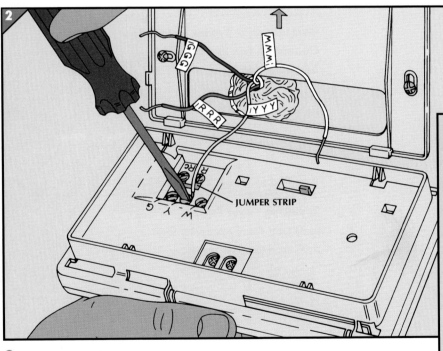

JUMPER STRIP

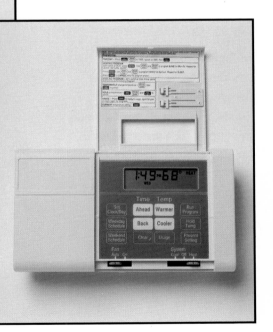

## 2. Wiring the thermostat.

◆ Holding the new thermostat next to the base plate, check that the wires are long enough to reach the new terminals. If they are not, cut 6-inch jumpers from 18-gauge thermostat wire. Strip insulation from both ends of the jumpers and join them to the existing wires with wire caps.

◆ Loosen all the thermostat terminal screws. Holding the unit in one hand, use the other to attach each coded wire to the corresponding terminal, then tighten the screw. If there is a fifth wire instead of four as shown here, break off the factory-installed jumper strip between the R and Rc terminals with a pair of long-nose pliers.

◆ Tuck in the wires and snap the thermostat onto the base plate. Install batteries and program the thermostat according to the manufacturer's instructions.

# The Most Heat for the Least Fuel

Furnaces and their system of pipes and ducts look forbidding, yet much furnace maintenance is surprisingly easy, and doing it yourself saves more than the cost of a service call. Some repairs and modifications to a system—adding a humidifier or electronic filter, for instance—can also be undertaken with basic tools and the techniques shown on the following pages.

Measuring the temperature rise in a forced-air system →

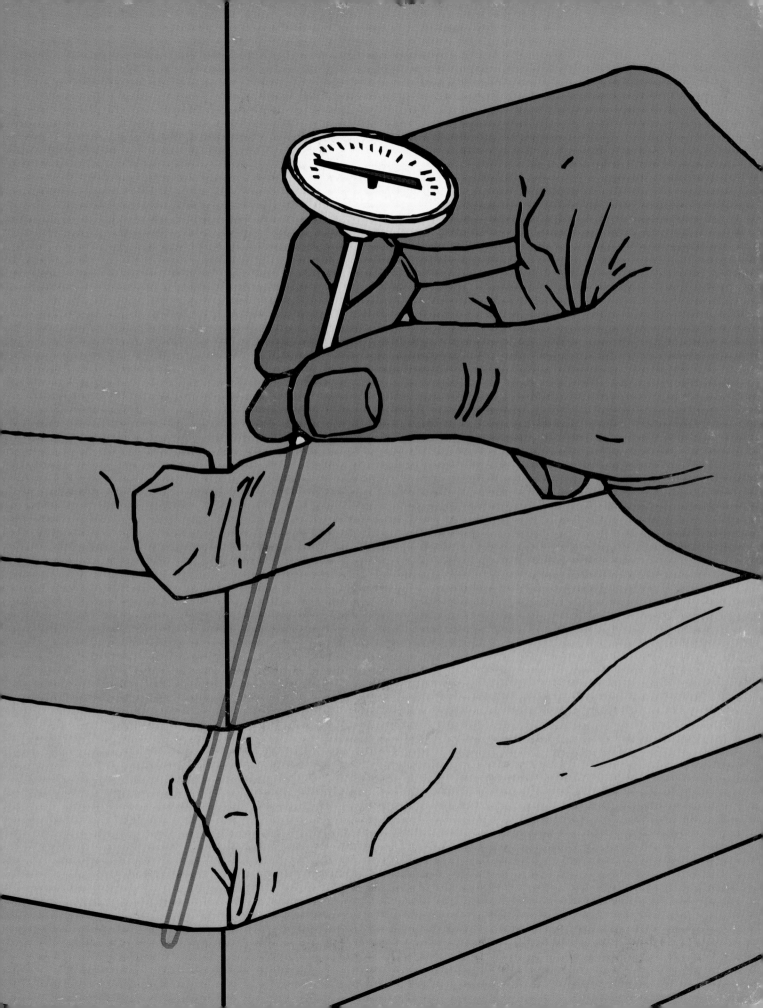

Gas furnaces, although extraordinarily reliable, require upkeep and occasional repair. Although many operations are best left to a professional, the ones shown in the chart at right—and explained on the following pages—are safe and uncomplicated.

**A Constant Pilot:** Originally, all gas furnaces had standing-pilot ignition—a small flame that lights the burners *(below, right)*. On this type of furnace, maintenance involves turning off the pilot in the spring to save energy and relighting it in the fall. Occasionally, the thermocouple—a sensor that detects whether the pilot is lit—may require replacement and the pilot and burner flames may need adjustment.

**Electronic Ignition Systems:** Instead of a standing pilot, new furnaces are likely to have an intermittent pilot—in which the flame is lit by a spark and burns only long enough to light the furnace—or by a heating element called a hot-surface igniter. If an intermittent-pilot system fails to spark, call for service. However, you can test and replace a hot-surface igniter that does not glow when the heating cycle starts *(page 21)*.

A periodic task necessary for so-called mid- and high-efficiency furnaces is checking the temperature rise—the difference between the air temperature in the supply and return ducts. Too high a difference may crack the heat exchanger, which transfers furnace heat to the air; condensation may corrode the system if the difference is too low.

**Diagnosing a Problem:** Furnaces with electronic ignition are regulated by a control center rather than by a series of relays and switches. Many centers have a red light that flashes a fault code to help in the diagnosis of problems. If the control center itself is faulty, it is easily replaced *(page 21)*.

⚠ **CAUTION** *If you smell gas, turn off the supply at the meter, ventilate the room, and do not touch electrical outlets or switches. If the odor persists, evacuate the house and call the gas company.*

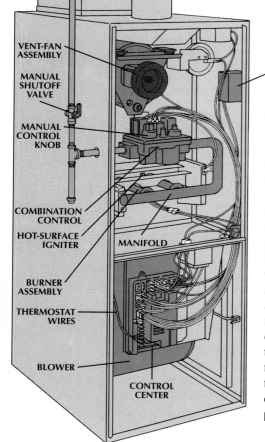

**VENT-FAN ASSEMBLY**

**MANUAL SHUTOFF VALVE**

**MANUAL CONTROL KNOB**

**COMBINATION CONTROL**

**HOT-SURFACE IGNITER**

**MANIFOLD**

**BURNER ASSEMBLY**

**THERMOSTAT WIRES**

**BLOWER**

**CONTROL CENTER**

**JUNCTION BOX**

**Two gas furnaces.**
With the burner access and blower compartment panels removed on these mid-efficiency *(left)* and standing-pilot *(right)* furnaces, their differences are readily apparent. Their basic operation is the same: A valve in the combination control releases gas, which travels through the manifold to burners, then a blower circulates warmed air through the house. Features such as a hot-surface igniter allow mid- and high-efficiency furnaces to make more heat with less fuel. A control center oversees lighting the burners and monitors safety features. It also regulates the vent-fan assembly, which draws off combustion by-products and disposes of them outside.

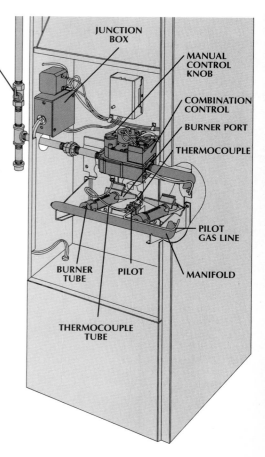

**JUNCTION BOX**

**MANUAL SHUTOFF VALVE**

**MANUAL CONTROL KNOB**

**COMBINATION CONTROL**

**BURNER PORT**

**THERMOCOUPLE**

**PILOT GAS LINE**

**MANIFOLD**

**BURNER TUBE**

**PILOT**

**THERMOCOUPLE TUBE**

**TOOLS**

 **MATERIALS**

 **SAFETY TIPS**

Screwdriver
Pocket thermometer
Nutdriver
Wrench
Hex wrench
Multitester

Hot-surface
  igniter
Control center
Vent-fan motor
Long match or
  soda straw

Thermocouple
  tube
Cloth

*When adjusting the air shutter on a burner, wear gloves to protect your hands from heat.*

---

## Troubleshooting Guide

| PROBLEM | REMEDY |
|---|---|
| **No heat.** | Replace fuse or reset circuit breaker.<br>Check control center for flashing fault code.<br>Replace hot-surface igniter *(page 21)*.<br>Replace vent-fan motor *(page 22)*.<br>Replace control center *(page 21)*.<br>Relight pilot *(page 23)*. |
| **Insufficient heat.** | Increase blower speed *(page 36)*.<br>Adjust burner air shutter *(page 25)*. |
| **Hot-surface igniter does not glow.** | Test and replace hot-surface igniter *(page 21)*.<br>Replace control center *(page 21)*. |
| **Pilot does not light or does not stay lit.** | Tighten or replace thermocouple *(pages 24-25)*. |
| **Pilot flame flickers.** | Adjust pilot flame *(pages 23-24)*. |
| **Exploding sound when burner ignites.** | Adjust pilot flame *(pages 23-24)*. |
| **Burner takes more than a few seconds to ignite.** | Adjust pilot flame *(pages 23-24)*. |
| **Burner flame too yellow.** | Adjust burner air shutter *(page 25)*.<br>Provide air from outside by opening vents in furnace room. |
| **Noisy furnace: rumbling when burners off.** | Adjust pilot flame *(pages 23-24)*. |
| **Noisy furnace: rumbling when burners on.** | Adjust burner air shutter *(page 25)*. |

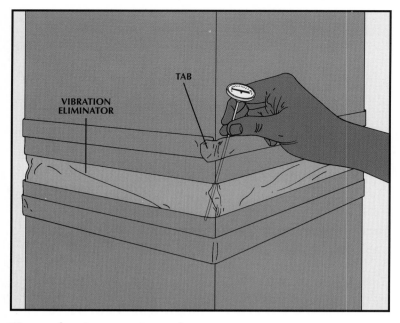

## Measuring temperature rise.

◆ With a screwdriver, loosen a slip and drive connector at a joint in the supply duct close to the furnace and insert a pocket thermometer *(above)*. (Alternatively, poke the thermometer through a canvas vibration eliminator between ducts; the hole will reseal itself.) Note the temperature.

◆ In the same manner, check the temperature in the return duct.

◆ Subtract the lower temperature from the higher one. If the difference falls below the range indicated on the information plate in the burner compartment, decrease blower speed *(page 36)*. For a result above the range, increase blower speed.

◆ If changing the blower speed does not rectify the problem, call a professional.

### Sequence of Furnace Operation

If you don't have a flashing fault code light to indicate where the heating cycle is interrupted, you can sometimes pinpoint a problem simply by watching and listening to learn where the furnace falters in its cycle, recounted here for a healthy mid-efficiency model.

✔ When the thermostat turns the furnace on, the vent-fan motor starts.

✔ After several seconds, the control center triggers the ignition system, causing the hot-surface igniter to glow or the spark igniter to flash.

✔ Gas is released from the combination control with an audible click. Normally the burners light.

✔ The flame sensor confirms that the burners have fired and allows the flow of gas to continue.

✔ When the furnace has warmed up, a temperature-sensitive switch or a timer turns on the blower, circulating air through the ducts.

# INSTALLING A HOT-SURFACE IGNITER

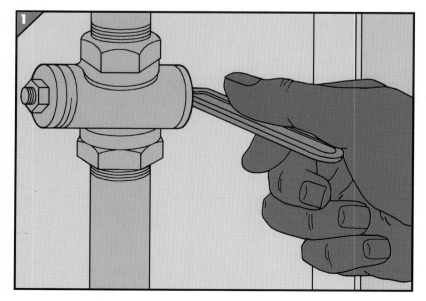

## 1. Shutting down the furnace.

◆ Turn off the gas to the furnace by closing the manual shutoff valve. The valve is closed when the handle is at right angles to the gas line *(left)*.

◆ Turn off power to the furnace at the master switch located on or near the furnace or, if there is no master switch, at the service panel.

## 2. Replacing the igniter.

◆ Take off the burner access panel.
◆ Unplug the igniter *(photograph)* and test it with a multitester *(page 118)*. Set the multitester to RX1 and attach an alligator clip to each prong in the igniter's plug. If the multitester registers between 45 ohms and 90 ohms, replace the control center *(below)*. Otherwise, install a new igniter.
◆ Unscrew or unclip the defective igniter and remove it.
◆ Screw or clip the new igniter in place; plug it in.
◆ Remount the access panel and restore power and gas to the furnace.

**CAUTION** *A hot-surface igniter is fragile and should be handled with care.*

# A NEW CONTROL CENTER

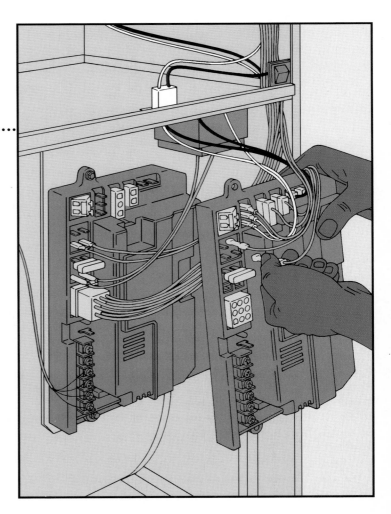

### Transferring the wires.
◆ Shut down the furnace *(page 20)*.
◆ Take off the blower access panel, removing the burner access panel first if necessary.
◆ With a nutdriver, remove the screws securing the control center. Leave the wires attached.
◆ Screw the new control center in place.
◆ Transfer the wires from the old center to the new, one at a time and in an orderly sequence.
◆ Remount access panels and restore power and gas to the furnace.

# REPLACING A VENT-FAN MOTOR

## 1. Removing the fan assembly.
◆ Shut down the furnace *(page 20)*.
◆ Remove the burner access panel and unplug the fan motor.
◆ Test the motor with a multitester *(page 118)* set to RX1. Attach an alligator clip to each prong on the plug. If the multitester registers any resistance at all, the motor is good; replace the control center *(page 21)*. Otherwise, install a new motor.
◆ With a nutdriver, remove the screws holding it in place. You may have to loosen the junction box or other components to reach all of the screws.
◆ Pull out the fan assembly.

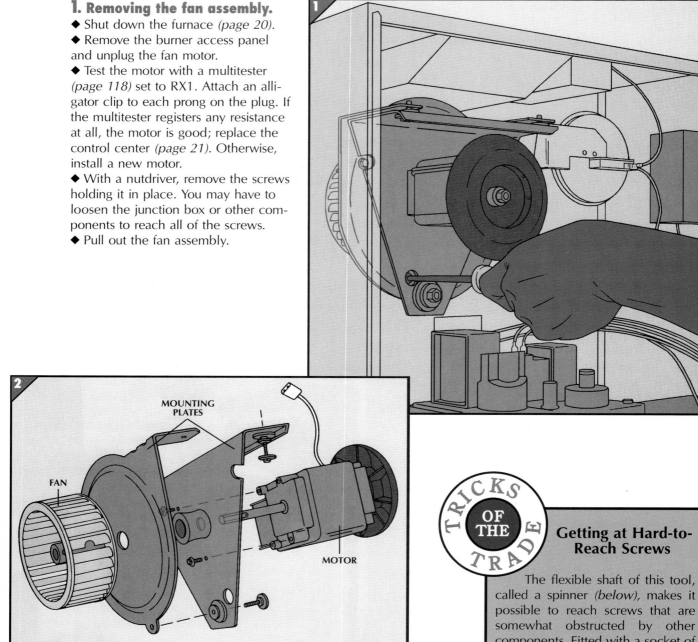

MOUNTING PLATES

FAN

MOTOR

## 2. Installing the new motor.
◆ Slip a hex wrench through the slot provided in the fan, loosen the setscrew, and pull the fan off the shaft. If the fan is damaged or rusty, buy a new one.
◆ Remove the screws holding the two mounting plates together and separate the plates.
◆ Unscrew the motor from its mounting plate.
◆ Fasten the new motor to the mounting plate, then screw the two plates together. Slip the fan wheel onto the motor shaft and tighten the setscrew.
◆ Install the assembly in the furnace and plug it in.
◆ Remount the access panel and restore power and gas to the furnace.

### TRICKS OF THE TRADE

### Getting at Hard-to-Reach Screws

The flexible shaft of this tool, called a spinner *(below)*, makes it possible to reach screws that are somewhat obstructed by other components. Fitted with a socket of the appropriate size, the shaft of the tool can bend slightly around an obstacle, often eliminating the need for disassembly.

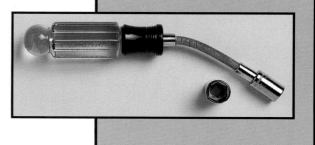

# LIGHTING A PILOT

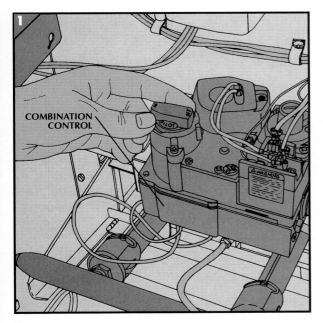

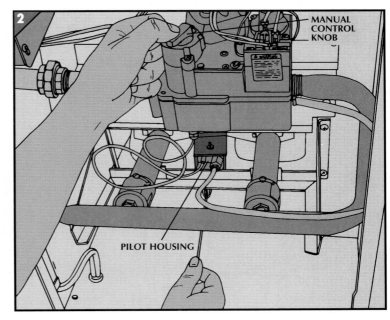

MANUAL CONTROL KNOB

COMBINATION CONTROL

PILOT HOUSING

### 1. Turning off gas to the pilot.
◆ Turn off power to the furnace at the master switch or control panel.
◆ Set the manual control knob, located on the combination control, to the off position *(above)* and wait 10 minutes for any gas to dissipate.

 **CAUTION** *If the smell of gas persists, do not attempt to relight the pilot; call for service.*

### 2. Relighting the pilot.
Follow the manufacturer's instructions for relighting the pilot—usually labeled on, or near, the combination control. In the absence of instructions, use the procedure explained here. If the pilot does not stay lit, the thermocouple may be faulty *(pages 24-25)*.
◆ Turn the manual control knob to the pilot position.
◆ While depressing the control knob or pilot ignition button, light the pilot burner located under the pilot housing with a long match *(above)* or a lit soda straw. Continue depressing the knob for 1 minute.
◆ Release the control knob, check the flame, and adjust as necessary *(Steps 1-2, below and page 24)*.
◆ Turn the control knob to the on position, then restore power to the furnace.

# ADJUSTING THE PILOT FLAME

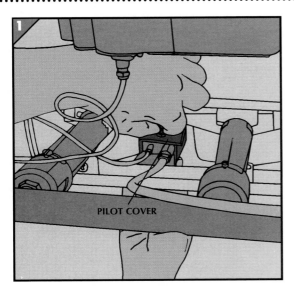

PILOT COVER

### 1. A view of the flame.
◆ Turn off power to the furnace at the master switch or control panel.
◆ If you can see the pilot flame clearly, go to Step 2. If, however, the pilot housing has a metal cover, first turn the manual control knob on the combination control to OFF. Wait a few minutes for the cover to cool. Unscrew and remove the cover *(left)*, then relight the pilot *(above)*.

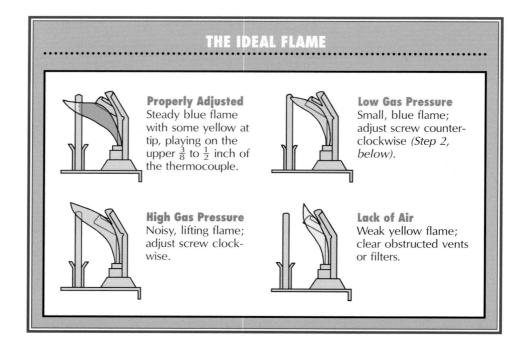

## THE IDEAL FLAME

**Properly Adjusted**
Steady blue flame with some yellow at tip, playing on the upper $\frac{3}{8}$ to $\frac{1}{2}$ inch of the thermocouple.

**Low Gas Pressure**
Small, blue flame; adjust screw counterclockwise *(Step 2, below).*

**High Gas Pressure**
Noisy, lifting flame; adjust screw clockwise.

**Lack of Air**
Weak yellow flame; clear obstructed vents or filters.

## 2. Adjusting the pilot screw.

◆ Find the pilot-adjustment screw on the combination control. On some models the adjustment screw is recessed and covered by a cap screw that must first be removed.

◆ To increase the height of the flame, turn the adjustment screw counterclockwise; lower the flame by turning the screw clockwise.

If you removed a pilot cover in Step 1, turn the manual control knob on the combination control to OFF, replace the cover, then relight the pilot *(page 23).*

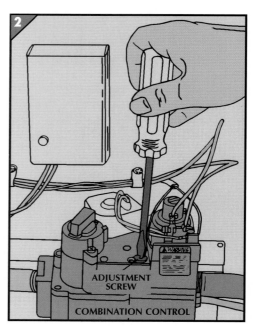

ADJUSTMENT SCREW

COMBINATION CONTROL

# EXCHANGING THE THERMOCOUPLE

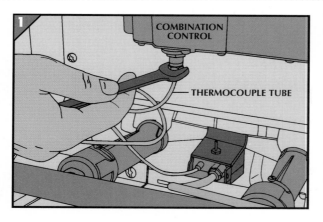

COMBINATION CONTROL

THERMOCOUPLE TUBE

## 1. Removing the thermocouple tube.

◆ First, shut down the furnace as described on page 20. A standing-pilot furnace may have a separate gas supply for the pilot; if so, also close the manual valve on this line. Allow a few minutes for metal parts to cool.

◆ Detach the thermocouple tube from the combination control by unscrewing the nut that secures the tube *(left).*

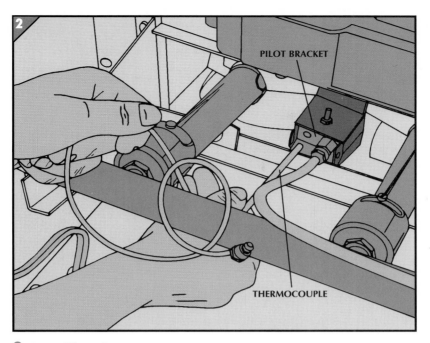

**2. Installing the new sensor.**

◆ Slide the defective thermocouple out of the bracket that holds it in place next to the pilot burner *(above)*.

◆ Use a cloth to clean the fitting on the combination control, and screw a new thermocouple tube into the fitting. After tightening the nut by hand, turn it a quarter-turn with a wrench.

◆ Being careful not to crimp the tubing, insert the thermocouple into the pilot bracket.

# OPTIMIZING A BURNER FLAME

**Rotating the air shutter.**

The burner may have an adjustable shutter, held in position by a lock screw on the end of the burner tube; or a fixed air shutter, which must be adjusted by a professional; or no air shutter at all. To adjust a movable shutter:

◆ Turn the thermostat to its highest setting to start the burner and keep it running. Allow 5 minutes for the burners to heat up, then remove the burner access panel and loosen the shutter lock screw *(inset)*.

◆ Slowly rotate the shutter open *(right)* until the blue base of the flame appears to lift slightly from the burner surface. Then close the shutter until the flame reseats itself on the surface. The flame should appear blue with a soft blue-green core and occasional yellow streaking. If not, call for service.

◆ Tighten the lock screw, then repeat the process for the remaining burners.

◆ Reset the thermostat.

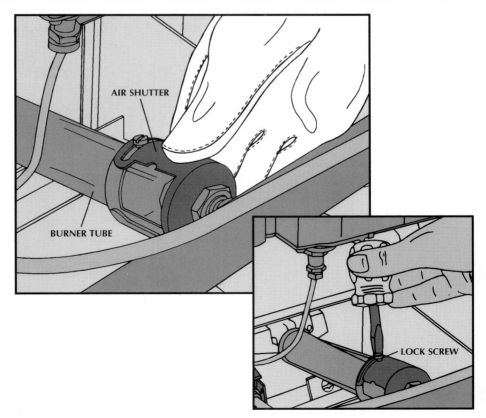

# Lining a Masonry Chimney

When vented through an unprotected masonry chimney, acidic combustion by-products released by modern gas furnaces can condense on the bricks or tile lining, causing gradual deterioration. An aluminum chimney liner solves the problem. You can install it yourself if your chimney flue is straight. Otherwise—or if you feel uncomfortable working on the roof or if it slopes more than 15 degrees—consider hiring a professional to do the work.

**Ordering the Liner:** The liner comes in a kit. Contents vary by manufacturer, but all contain an expandable liner, a mortar guard to protect it where it exits the chimney's bottom, and metal flashing and a rain cap for the top. Kits with no insulating sheath are simpler to install.

Before buying a liner, you need to know the height of your chimney, the interior dimensions of the flue inside it, and the lengths of the ducts—called vent pipes—from the furnace and water heater, if you have one, to the chimney opening.

Also check the labels on the appliances for their ratings in British thermal units per hour (BTUH) and for whether the furnace has a fan to assist in expelling exhaust gases. With this information a heating-supply store can direct you to the correct liner for your circumstances.

**Preparing the Chimney:** Check with local building or fire officials for venting restrictions and inspections required in your area. To remove soot and other debris from the chimney, pour 2 to 3 pounds of sand into

a cloth bag about half the size of a pillowcase. Tie a $\frac{1}{2}$-inch rope around the bag and lower it down the chimney several times, letting the bag rub against the sides of the flue.

**Connecting the Appliances:** If the existing vent pipe, which links the furnace to the chimney, is not corroded, disconnect it at the joint nearest the chimney and attach the liner directly to it. If venting a gas water heater too, you may have to reconfigure the vent pipes to maintain the proper upward slope—at least $\frac{1}{4}$ inch per foot—where the pipes connect to the T or Y. Some chimney liner kits require a single-wall pipe to be replaced by one with double walls, called a B-vent pipe, to prevent backdrafts that can trap hazardous exhaust gases in the house.

---

 **TOOLS**

Electric drill with $\frac{5}{16}$" magnetic nutdriver bit
Broad-billed pliers
Caulking gun
Utility knife
Tin snips
Trowel

 **MATERIALS**

Rope ($\frac{1}{2}$")
Cloth bag
Sand (2 to 3 lb.)
Silicone sealant
Self-tapping sheet-metal screws (No. 10, $\frac{3}{4}$")
Mortar
Vent T, single- or double-wall

**SAFETY TIPS**

*Wear work gloves when handling metal flashing, ductwork, or wet mortar. Protect your eyes with goggles when drilling at or above eye level, and wear nonslip rubber-soled shoes when working on the roof.*

---

### 1. Stretching the liner.

◆ Shut off the gas and power to the furnace and any other appliance you plan to vent through the liner *(page 20)*.

◆ With a helper, gently stretch the compressed chimney liner to its full length. Beginning near the middle *(left)*, stretch one end completely, small sections at a time, then have your helper expand the other end.

⚠️ **CAUTION** *Chimney liners are fragile. To avoid tears, stop stretching before all the corrugations flatten out, and take care not to collapse the liner with too firm a grip.*

## 2. Inserting the liner.

◆ Tie a rope to the end of the liner and lift it to the roof, being careful not to tear it.

◆ Slide the liner down the chimney *(above)* while your helper guides the liner through the hole at the base of the chimney. Leave 6 to 12 inches of liner extending from the top of the chimney.

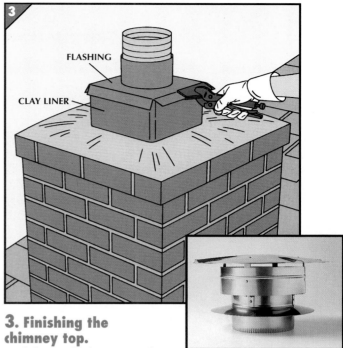

## 3. Finishing the chimney top.

◆ Slide the flashing over the end of the liner and center it over the chimney opening. If there is a clay chimney liner *(above)*, or if the flashing extends beyond the sides of the chimney, cut out the corners and bend the edges down with broad-billed pliers.

◆ To secure the flashing, spread silicone sealant under the edges, then caulk around them with the sealant.

◆ With a knife or tin snips, trim the liner 1 inch above the chimney flashing, then slip the rain cap *(photograph)* into the top of the liner and attach it with three evenly spaced self-tapping sheet-metal screws.

◆ Have a helper pull on the bottom of the liner to lower the rain cap onto the top of the flashing.

## 4. Connecting the liner to the furnace.

◆ At the base of the chimney, slide the mortar guard onto the liner and into the hole in the chimney. Install the mortar guard so that the Underwriters Laboratory label *(not shown)* is visible for future reference. Pack mortar evenly around the mortar guard with a small trowel.

◆ Trim the liner about a foot from the mortar guard and attach the furnace vent pipe to the liner with self-tapping sheet-metal screws.

To vent a water heater, fit a T or Y between the furnace vent pipe and the liner *(inset)*, then fasten the water heater vent pipe to the branch of the T or Y. Patch the old water heater vent hole in the chimney.

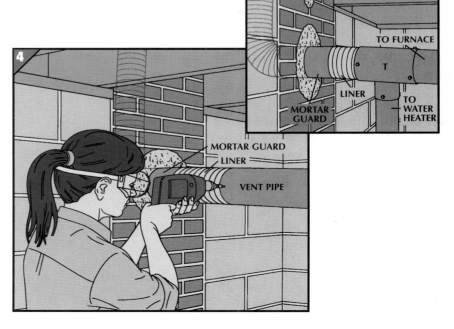

# Tuning Up an Oil Burner

Cleaning and adjusting an oil burner before the heating season starts can cut fuel bills and extend the life of the burner. Moreover, the trouble-shooting chart opposite shows which of the procedures on the following pages can prevent costly service calls. Once a year, however, have a professional check the efficiency of the burner, a job that requires costly test instruments.

**Before Beginning Work:** Use newspapers to protect the floor, a pan of sand or cat litter to catch drips, and a bucket to dispose of sludge and excess oil. Turn off the burner master switch and cut power to the circuit that governs the burner by switching off the circuit breaker or pulling the fuse. Then shut down the oil line at the valve between the filter and the storage tank.

If your oil line has a special fire safety valve at the pump, turn the handle clockwise to push the stem down. The valve is closed when the handle slips off the stem. To seat the valve completely, give the stem a light tap with a wrench.

When the job is done, reopen the oil line before restoring power to the burner. Doing so prevents air pockets in the oil supply.

**Repairing the Furnace:** An air leak, a dirty heat exchanger, soot in the flue and the chimney, or a crumbling combustion chamber all affect burner efficiency. You can seal leaks with furnace cement *(page 34)* and clean heat exchanger or boiler surfaces with a stiff brush and a vacuum cleaner.

A bag of sand loosens soot, removable through a cleanout door at the base of the chimney. To dislodge soot in the stack leading from the furnace to the chimney, dismantle it and rap each section against a floor covered with newspapers.

**Renewing a Combustion Chamber:** After much intense heat and expansion and contraction, a combustion chamber's lining may crumble or burn up. If a firebrick lining crumbles in large chunks, furnace cement can bond the pieces back in place.

But the best repair for any combustion chamber retaining its shape is to reline it. Available from heating and refrigeration suppliers, liners are made of heat-resistant fibers, kept moist and flexible in a plastic bag until heat is applied after installation. Before buying one, make sure the modification will not void your furnace or boiler warranty.

To get the right liner, you will need the firing rate and the spray pattern of your nozzle *(page 32, Step 3)* and the dimensions of the combustion chamber *(page 35, Step 2)*. The dealer may suggest a different nozzle, one more compatible with the new liner.

**An Uncontaminated Oil Tank:** Water or rust in the fuel line can cause burner problems. To keep moisture from condensing inside the oil tank and rusting the bottom, fill the tank at the end of the season.

You can detect water in the tank with gray litmus paste (from a heating or plumbing supplier) smeared on the bottom of the dipstick used to gauge fuel level. If the paste turns purple, water has pooled in the bottom of the tank. Drain water more than an inch deep from the bottom of the tank.

 **TOOLS**

Screwdriver
Shallow pan
Bucket
Small toothbrush
Long, narrow brush
Wrenches
Wire brush
Putty knife
Tape measure
Scissors

 **MATERIALS**

Cat litter or sand
Oil filter cartridge
Filter-bowl gasket
Nondetergent
    electric-motor oil
Clean rags
Solvent
Stiff paper
Furnace cement
Combustion-
    chamber liner

## How an oil burner works.

When the thermostat calls for heat, the burner's relay box, or primary control, turns on the burner motor, which pumps oil to the burner nozzle and blows air into the combustion chamber. The pump draws oil from a storage tank through a shutoff valve and filter, then blasts it out the nozzle in a fine mist that mixes with blower air entering the combustion chamber through the burner air tube. Simultaneously, an ignition transformer boosts household voltage from 120 volts to 10,000 volts and sends it to the electrodes, causing a spark that ignites the oil-air mixture. Combustion gases exit through a stack at the back of the furnace *(not shown)*. In the event of a misfire, a light-sensitive photoelectric cell shuts down the system until a reset button on the relay box is pressed.

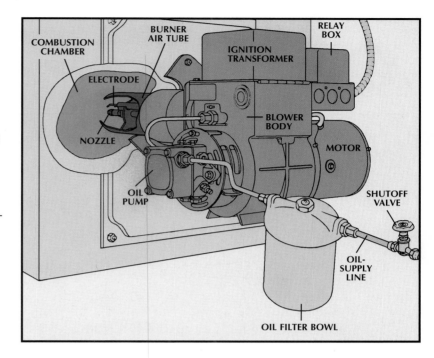

## Troubleshooting Guide

| PROBLEM | REMEDY |
| --- | --- |
| **No heat or insufficient heat.** | Check oil level in the fuel tank; have it refilled if necessary. |
| | Press reset button. (If the system doesn't start after two attempts, call for service.) |
| | Check electrodes; adjust if necessary. |
| | Clean or replace nozzle *(pages 32-33)*. |
| | Clean photocell *(page 31)*; replace if necessary. |
| | Replace oil filter *(page 30)*. |
| | Clean strainer *(page 30)*; replace if necessary. |
| **Intermittent heat.** | Check oil level in the fuel tank; have it refilled if necessary. |
| | Press reset button. (If the system doesn't start after two attempts, call for service.) |
| | Replace oil filter *(page 30)*. |
| | Clean strainer *(page 30)*; replace if necessary. |
| | Replace nozzle *(pages 32-33)*. |
| | Clean or replace photocell *(page 31)*. |
| **High fuel consumption.** | Replace nozzle *(pages 32-33)*. |
| **Burner system noisy.** | Lubricate motor *(page 31)*. |
| **Diesel fuel odor from burner system.** | Tighten oil line fittings. |
| | Check electrodes *(page 32)*; adjust if necessary. |

# UNCLOGGING THE FILTER SYSTEM

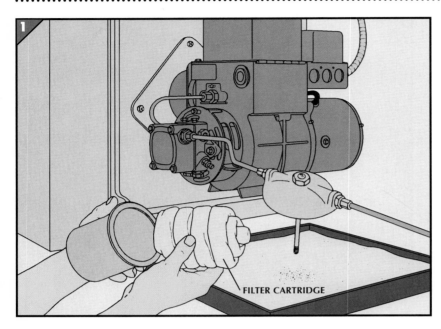

FILTER CARTRIDGE

## 1. Changing the oil filter.
◆ Switch off the burner, shut off the burner circuit, and close the supply valve. Set a pan filled with cat litter or sand under the oil filter.
◆ Unscrew the bolt on the cover above the filter bowl. Remove the bowl and upend it into a bucket, letting the filter cartridge inside fall out.
◆ Wipe the bowl and peel off the old gasket.
◆ Insert a new cartridge and place a new gasket on the lip of the filter bowl, then reattach the bowl to the cover.

## 2. Cleaning the pump strainer.
If your pump has a rotary-blade filter instead of a strainer—your oil dealer can tell you—skip this step. Otherwise, proceed as follows:
◆ Unbolt the pump cover and set it aside, without disconnecting the oil line. Discard the thin gasket around the cover rim.
◆ Remove the cylindrical wire-mesh strainer. Replace a torn or bent strainer; soak one that passes inspection in solvent for a few minutes to loosen sludge buildup. Then clean the mesh gently with an old toothbrush.
◆ Reinstall the strainer, place a new gasket on the cover rim, and bolt the cover in place.

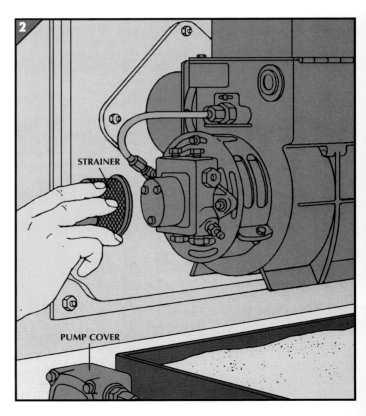

STRAINER

PUMP COVER

# CHECKING THE FAN AND MOTOR

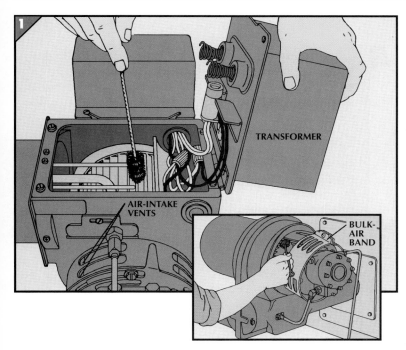

## 1. Cleaning the fan.

To maintain burner efficiency, clean the fan and air-intake openings every month during the winter as follows:

◆ First, sweep out the air-intake vents on the fan housing with a long, narrow brush.

◆ Unscrew the transformer atop the burner and swing it out of the way to expose the fan. Brush the fan blades *(left)* and wipe the interior of the fan housing with a rag.

To reach the fan on an old-style burner *(inset),* mark the position of the slotted bulk-air band that surrounds the housing. Loosen the screw holding the band and slide the band back. After cleaning the fan, reposition the band in its original location and tighten the screw.

## 2. Lubricating the motor.

◆ On a burner motor with small oil cups at each end—the absence of oil cups indicates a permanently lubricated motor—lift the lids or plugs from the cups.

◆ Dribble 4 or 5 drops of 10- to 20-weight nondetergent electric-motor oil in each cup and replace the lids or plugs.

◆ Lubricate the motor every 2 months, or at the intervals specified by the manufacturer.

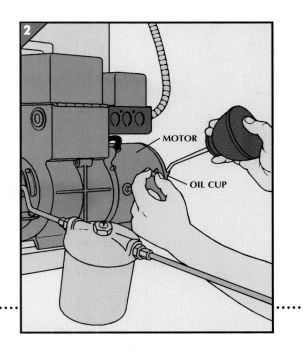

# CLEANING THE FLAME SENSOR

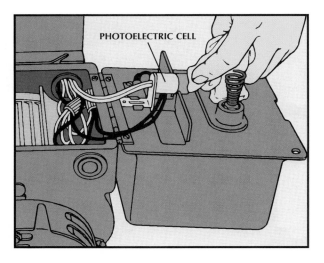

## The light-detecting cell.

◆ Unscrew the transformer atop the burner and lift it up. Most often, the photoelectric cell that shuts off the motor when ignition fails is mounted on the underside of the transformer or attached to the burner housing near the end of the air tube.

◆ Wipe dirt from the cell with a clean rag, then resecure the transformer.

# MAINTAINING THE FIRING ASSEMBLY

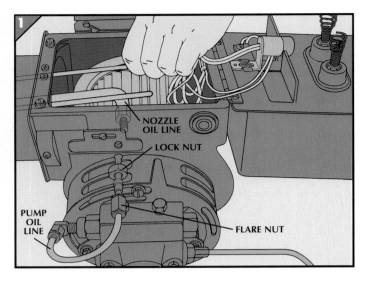

## 1. Removing the firing assembly.

◆ After moving the transformer out of the way, mark the position of the firing assembly in the air tube.

◆ Disconnect the nozzle oil line from the pump oil line, first loosening the flare nut then the lock nut.

◆ Pull the entire firing assembly—electrodes and nozzle oil line—out of the air tube.

You may need to twist the assembly as you pull it, but be careful not to knock the electrodes or nozzle against the burner housing.

◆ With the firing assembly removed, clean the air tube with a cloth or a brush. If there is a flame-retention device—a circular metal piece with fins or vanes—at the end of the tube, clean it as well.

## 2. Cleaning the ignition system.

◆ With a cloth dipped in solvent, wipe soot off the electrodes and their insulators, as well as the electrode extension rods or cables and the transformer terminals.

◆ If the insulators are cracked or the cables frayed, take the entire assembly to a professional for repair.

◆ Measure the spacing of the electrode tips, which should exactly match the manufacturer's specification—usually about $\frac{1}{8}$ inch apart pointed toward each other, no more than $\frac{1}{2}$ inch above the center of the nozzle tip, and no more than $\frac{1}{8}$ inch beyond the front of the nozzle (inset). If necessary, loosen the screw on the electrode holder and gently move the electrodes into place.

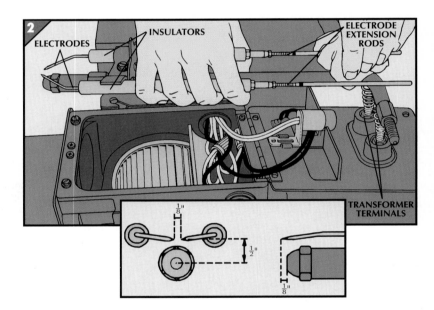

## 3. Removing the nozzle.

◆ With one wrench holding the hexagonal adapter at the end of the nozzle oil line, unscrew the nozzle with another. Take care not to twist the oil line or alter the positions of the electrodes.

◆ Examine the tip of the nozzle (inset); the stamped specifications show the firing rate in gallons of oil per hour, or gph (1.75 in this case), and the angle of spray (60 degrees). Letters usually identify the type of spray pattern.

◆ If the nozzle has a firing rate of 1.50 gph or less, replace it with an identical nozzle. If the nozzle has a higher rate, you can clean and reuse it (Step 4).

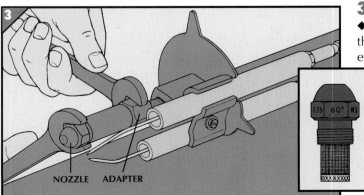

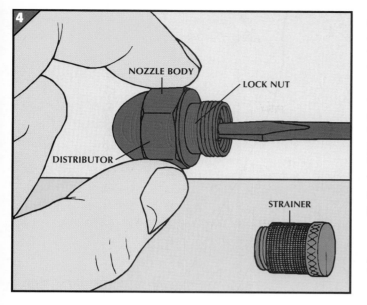

## 4. Disassembling the nozzle.

◆ With your fingers, unscrew the strainer from the back of the nozzle. Then unscrew the lock nut holding the distributor inside the nozzle tip *(left)*. Slide the lock nut and distributor out of the nozzle body.

◆ Soak all nozzle parts in solvent for a few minutes, then scrub gently with a small toothbrush. Clean distributor slots with a piece of stiff paper, and the nozzle orifice with compressed air or a clean bristle. Flush all the parts with hot water, shake them, and air-dry them on a clean surface.

◆ Reassemble on a clean surface, making sure your hands and tools are clean. Screw the nozzle onto the adapter finger tight, then snug it a quarter-turn with the wrenches.

 **CAUTION** *Never use a pin or wire to clean the nozzle orifice; scratches might alter the spray pattern.*

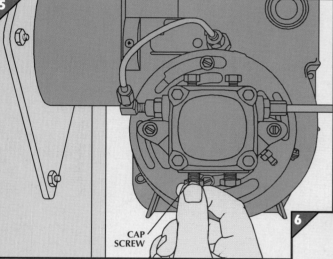

## 5. Opening the oil line.

(Skip this step if your storage tank is below the pump on your oil burner.)

◆ Loosen the cap screw on the unused intake port of the underside of the pump cover *(left)*.

◆ Place the pan filled with cat litter or sand under the oil pump and open the supply valve. When oil begins to flow from the intake port, let it run into the pan for about 15 seconds before tightening the cap screw.

## 6. Priming the pump.

◆ Loosen the pump oil line at the pump to feed the unattached end into a bucket. (For safety, temporarily swing the transformer down or, on an old-style burner, screw on the rear plate.) Set the house thermostat to a high temperature and restore power at the service panel.

◆ Hold the bucket and have a helper throw the master switch. Oil will gush from the line with great force. Let the pump run for about 10 seconds, then have your helper turn off power at the master switch and the service panel.

◆ Lift the transformer (or remove the rear plate). Guided by the marks made in Step 1, install the firing assembly in the air tube, centering the nozzle oil line in the tube. Connect the pump and nozzle oil lines by tightening the lock nut and flare nut with your fingers—plus a quarter-turn with a wrench.

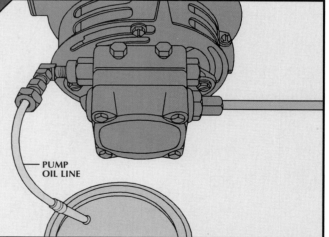

◆ Screw the transformer down (on an old-style burner, reconnect the electrodes and replace the rear plate). Restore power at the service panel.

◆ To expel any air that may remain in the oil line, partially open the observation port in the fire door and turn on the burner at the master switch. Run the burner for 10 seconds and shut it off. Repeat five times, or until the burner shuts down smoothly and instantaneously.

# MAKING AN OIL BURNER AIRTIGHT

### 1. Locating leaks.
◆ Examine the flue and replace any section that is badly rusted and perforated with small holes.
◆ Check for leaks at the seams by firing up the burner and moving a lighted candle along each seam. The flame will deflect inward at leaks. Use this method to inspect the combustion-chamber cover plate, the burner mounting flange, the fire door, and the flue joints *(red)*.

### 2. Sealing a leak.
◆ Turn off the burner and allow the furnace to cool. Clean the surfaces around the leak with a wire brush. Use a putty knife to fill gaps with refractory furnace cement.
◆ To seal a leak around the burner mounting flange *(right),* loosen the bolts around the edge and pull the flange back a fraction of an inch. Scrape away old gasket material under the flange and apply a thin layer of cement around the edges. Retighten the bolts.

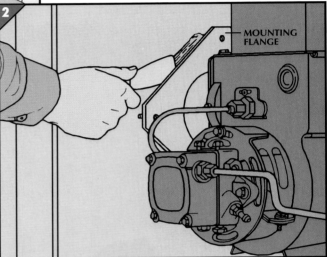

# RELINING THE COMBUSTION CHAMBER

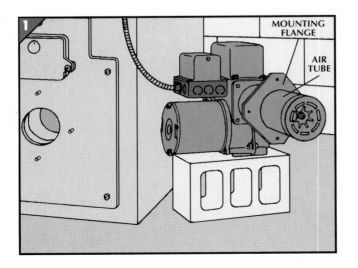

### 1. Removing the burner.
◆ Shut down the burner at the master switch and service panel, then close the oil-supply valve. Mark the air tube so that you can insert it the same distance into the combustion chamber when you reinstall the burner.
◆ Unscrew the bolts on the mounting flange and pull the burner away from the combustion chamber. If you cannot do so without bending the oil-supply line, disconnect the line at the oil-burner pump.
◆ As you pull the burner from the air-tube port, make sure that any gasket material encircling the air tube does not fall off and break. Set the burner down on its own pedestal or support it on a cinder block.

## 2. Measuring the combustion chamber.

◆ Inspect the chamber by looking through the air-tube port or fire door and by reaching inside and feeling the walls and floor with your hand.
◆ Measure the depth of the chamber with a tape measure or yardstick *(right)*.
◆ Calculate chamber width by measuring the width of the furnace and subtracting twice the thickness of the combustion-chamber walls, measured at the air-tube port.
◆ Take your measurements to a heating supplier and buy a liner to fit.

⚠ **CAUTION** *In old oil burners, the combustion liners and the patching material used around joints (opposite) may contain asbestos. Before disturbing any suspect material, test it (page 48).*

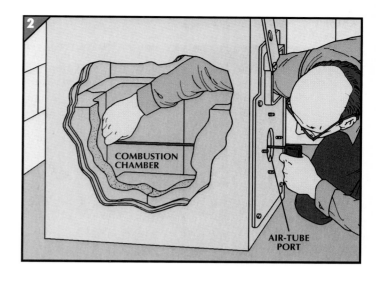

## 3. Preparing the liner.

◆ Remove the damp liner from its plastic bag and spread it open.
◆ Measure the height of the combustion chamber and mark it on the liner. Use scissors to cut from the liner edge to the line to make flaps 4 inches wide.

◆ If the liner does not have a hole for the air tube, measure from the top of the combustion chamber to the top of the air-tube port. Then, the same distance below the flaps, cut out a circle slightly smaller than the diameter of the air tube.

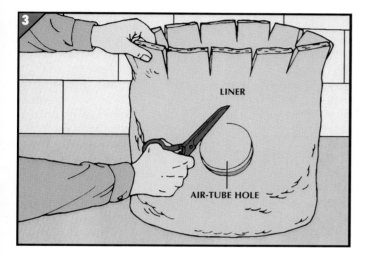

## 4. Lining the chamber.

◆ Roll up the liner and push it into the combustion chamber through the air-tube port. Reach through the fire door and the air-tube port and unroll the liner; align the air-tube hole with the air-tube port.
◆ Working back to front, mold the liner against the walls and floor of the chamber. Pat the liner smooth and fold any flaps at the top of the liner over the top edge of the chamber.
◆ If the liner tears, press the torn edges back together. You can also patch torn sections with scraps from the air-tube hole.
◆ Partially dry the liner with a light bulb of at least 100 watts until it has the consis-

tency of stale bread. Then trim the air-tube opening in the liner with a sharp knife so that the edge is flush with the air-tube port.
◆ Push the air tube into the port up to the mark made in Step 1. Screw the mounting flange to the furnace. Reconnect the oil line, if you disconnected it earlier, and prime the pump *(page 33, Step 6)*. In all cases, open the oil valve and restore power to the burner.
◆ Turn on the burner at the master switch, let it run for 3 minutes, and shut it off for 3 minutes. You may see a little smoke and detect an unfamiliar odor; both are normal. Repeat this procedure twice to set the liner.

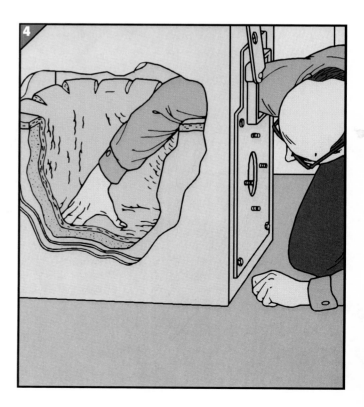

The blower—or fan—that distributes the air in a typical forced-air system is spun by a motor attached to the fan shaft. Such a unit seldom needs repairs, and any problems that do occur are generally easy to correct.

**Poor Air Flow:** If the blower fails to deliver enough warm or cool air, dust and lint may be clogging the system. Check the filter to see if it needs to be cleaned or replaced. At the same time, pull the blower from the unit *(opposite, top)*, brush any dirt from the fins of the blower wheel, then vacuum it out.

Proper air flow also depends on blower speed. To strengthen or moderate the flow on older units, adjust the speed *(below)*. Before changing the blower speed on mid- and high-efficiency systems—which came into use in the late 1980s—read the information on checking the temperature rise *(page 20)*.

**Noise:** Vibration noises often can be quieted simply by tightening the screws holding the blower housing and the motor. A squealing or grating noise may be due to dry bearings in the blower: Although newer

fans have permanently lubricated and sealed bearings, some older models require oiling before the heating season starts, and again at the end of the season if the blower also serves a central air conditioner.

**Replacing the Motor:** If a blower motor burns out, replace it with one of the same size *(opposite, bottom)*.

⚠️ **CAUTION** *Before you begin any job that involves touching the unit's wires, turn off the master switch, and also shut off the circuit breaker or remove the fuse.*

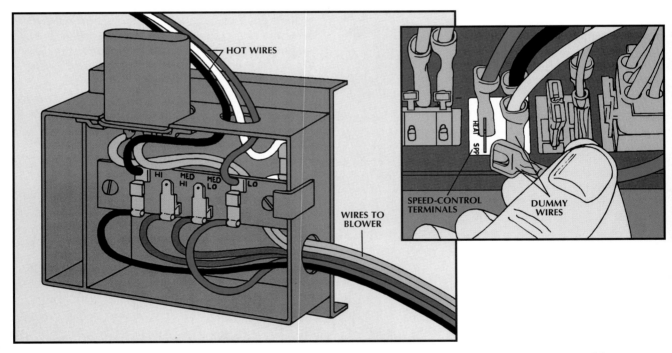

HOT WIRES

HI  MED HI  MED LO  LO

WIRES TO BLOWER

SPEED-CONTROL TERMINALS

DUMMY WIRES

## Adjusting a multispeed motor.

An older unit may have a blower-mounted junction box *(above)* housing the blower-speed terminals. More modern units may have a set of plastic modular plugs *(inset)* mounted on the blower. In both cases the blower speed is adjusted by changing the positions of the hot wires. A hot red wire connects to the low or medium-low terminal to run the blower when the furnace is in

operation. A hot black wire connects to the high or medium-high terminal to supply power during air conditioning. The unit also may have blue or yellow "dummy" wires that carry no electrical power *(inset)* on unused terminals. To change the blower speed proceed as follows:

◆ Remove the access panels from the unit—and from the junction box on older models.

◆ To increase or decrease blower speed for the heating system, unplug the red hot wire and attach it to the adjacent terminal.

◆ To increase or decrease blower speed for the air conditioner, unplug the black hot wire and attach it to the adjacent terminal.

◆ If the panel has dummy wires, exchange the position of one with that of the red or black wire *(inset)*.

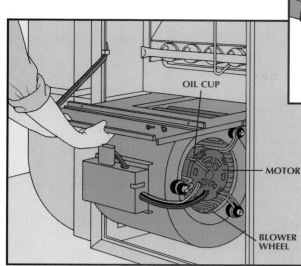

OIL CUP

MOTOR

BLOWER WHEEL

## Oiling the bearings.

◆ Remove the access panels from both the blower and furnace compartments. If the control panel is in the way, remove the screws or nuts holding it in place, and move the panel aside.

◆ Loosen the screws or nuts securing the metal blower shelf to the unit.

◆ Slide the blower partway forward by pulling the shelf. If the electrical wires are too taut to permit the blower to slide easily, unclip the wires from the side of the furnace or detach them at the blower-speed terminals (opposite page, bottom).

◆ Look for oil cups (inset) or plastic plugs at the visible end of the blower motor; if there are none, lubrication is not needed.

◆ Otherwise, lift the lids of the cups or pull the plugs and drip six to eight drops of 10- to 20-weight nondetergent electric-motor oil into each.

◆ Slide the blower back into place, replace the screws or nuts, and put back the access panels.

TRICKS OF THE TRADE

### Oiling from a Distance

The oil ports on some motors are deep in the blower and hard to reach. To get lubrication to an inaccessible oil port, insert a thin wire into the port, hold the wire in a vertical position, and drip oil slowly onto the wire so that it runs into the hole.

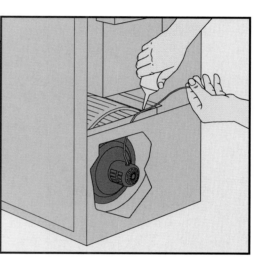

## Replacing the motor.

◆ Unplug the blower wires from the blower-speed terminals (opposite, bottom).

◆ Slide the blower and attached shelf onto the floor.

◆ Loosen the bolt that connects the end of the motor shaft to the blower wheel.

◆ Remove the bolt assemblies that are holding the motor mounting bracket to the blower housing; ease the motor out of the blower.

◆ Remove the nut and washer at the ends of the mounting bracket's braceletlike ring (inset).

◆ Slip the bracket off the motor and attach it to the new motor in approximately the same position as it was on the old one.

◆ Slide the motor into the blower, reattach the mounting bracket, and tighten the bolt against the flat spot on the motor shaft.

◆ Rotate the blower wheel by hand; if the wheel rubs the housing, loosen the bolt and shift the wheel sideways until it is able to rotate freely.

◆ Slide the blower back in place.

◆ Tighten the screws or nuts, reconnect the wires, and replace the access panel.

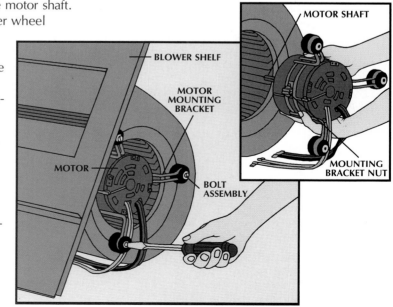

BLOWER SHELF

MOTOR MOUNTING BRACKET

MOTOR

BOLT ASSEMBLY

MOTOR SHAFT

MOUNTING BRACKET NUT

During the heating season, the air in a house can become uncomfortably dry. To solve this problem, homes with forced-air heating systems often have a central humidifier to maintain a constant level of 30 to 35 percent relative humidity. If your furnace already has a humidifier, the simple maintenance steps shown here will keep it running smoothly. Or you can add a humidifier to your system *(pages 41-43)*.

**Bypass or Fan-Powered:** Humidifiers, all of which transfer moisture to air circulating through a water-soaked pad, are available in two types. Bypass humidifiers use the difference in air pressure between the plenum and the return-air duct to force air through a rotating drum pad. Fan-powered humidifiers have a motor that blows air through a flat pad and directly into the plenum.

**Routine Upkeep:** Bypass humidifiers need cleaning *(Steps 1-3, below and opposite)* more than once a season to prevent mineral buildup in the drum-shaped pad. How often to clean depends on the hardness of the water in your area. Fan-powered units are less trouble; cleaning their flat pads once a year will usually be sufficient. Moreover, because they don't have a water reservoir, you never have to clean the tray or adjust the water level.

**Installing a Humidifier:** Since fan-powered humidifiers require less upkeep than bypass models—and because they use less water—they are also the kind most often added to a heating system. In addition to mounting the humidifier on the plenum, the installation calls for a humidistat—a sensor that controls the level of humidity. Wiring for a 120-volt model is shown on the following pages. Hooking up a 24-volt unit is much the same, except that power comes from the furnace transformer and the color coding of the wires will differ.

**TOOLS**
Wrench
Screwdriver
Carpenter's level
Electric drill
Awl
Tin snips

**MATERIALS**
Saddle valve
2-wire grounded cable
   (No. 14)
Wire caps

**SAFETY TIPS**
*Wear goggles and gloves when cutting sheet metal.*

## MAINTAINING A BYPASS HUMIDIFIER

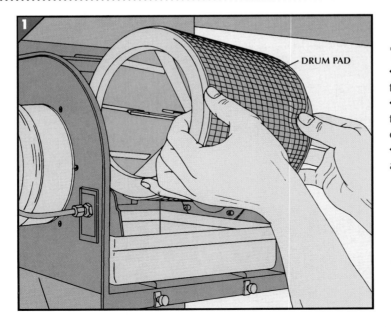

DRUM PAD

**1. Removing the drum pad.**
◆ Turn off power to the furnace at the master switch or service panel.
◆ Loosen the retaining nuts along the bottom lip of the humidifier cover and lift it off.
◆ Grasp both ends of the drum and lift it out of its slots.

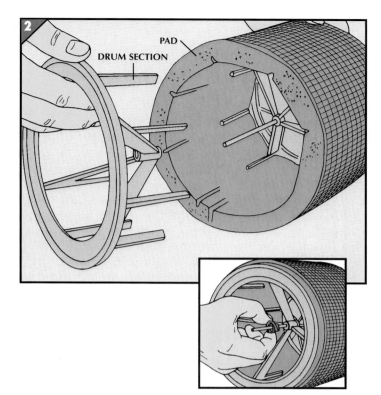

## 2. Cleaning the pad.

◆ Pinch the drum-shaft retaining clip and pull it off the shaft *(inset)*. Then pull apart the two sections of the drum to release the pad *(left)*.

◆ If the pad is slightly hardened, soak it in a solution of 3 parts vinegar to 1 part water until it softens. Squeeze the solution through the pad to clean it thoroughly.

◆ Replace a pad that remains brittle even after soaking or one that breaks apart while cleaning.

## 3. Cleaning the tray.

◆ Turn off the water line valve, lift the float, and pull out the tray *(right)*.

◆ Using a stiff brush, scrub the tray with a vinegar-and-water solution or a commercial humidifier descaler to dislodge deposits. Rinse the tray well.

◆ Return the tray to the humidifier and turn on the water. Add water-treatment tablets or liquid—available at hardware stores—to inhibit mineral buildup and bacteria.

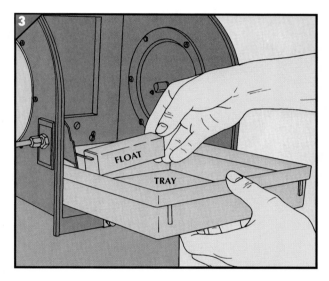

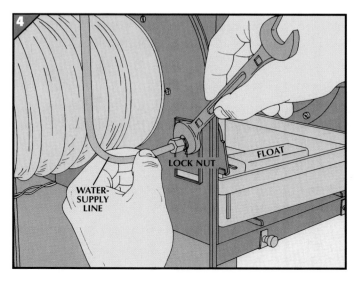

## 4. Adjusting the water level.

◆ Measure the depth of water in the tray; it should be about $1\frac{1}{2}$ inches deep, enough to soak the pad as it turns.

◆ To adjust the water level, loosen the float assembly lock nut on the water-supply line with a wrench *(left)*. Move the float up to raise the water level, or down to lower the water level.

◆ Retighten the lock nut securely against the retainer plate and recheck the new water level. Then replace the humidifier cover and restore power.

# CLEANING A FAN-POWERED MODEL

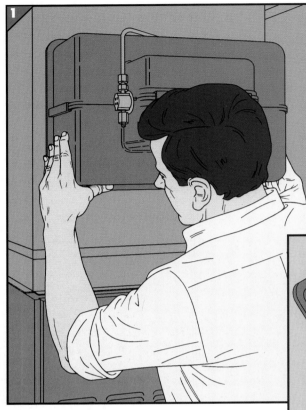

## 1. Removing the humidifier.
◆ Turn off power to the furnace at the master switch or service panel and shut off the water supply to the humidifier. Then, with a wrench, disconnect the line where it enters the humidifier.
◆ Unscrew the clamp on the plastic overflow tube and pull it free from the bottom of the unit.
◆ Grasp the humidifier at the lower two corners and push up *(left)*, then pull the humidifier away from the warm-air plenum and place it on the floor.
◆ Slide off the plastic clips that hold together the top and bottom halves of the humidifier cover *(inset)*. Lift off the top half of the humidifier cover.

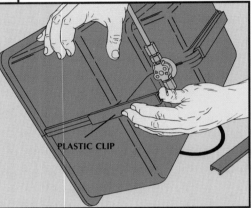

PLASTIC CLIP

## 2. Cleaning the pad.
◆ Lift the plastic pad assembly up and out of the humidifier.
◆ Slide the pad from the holder and soften it in a solution of 3 parts vinegar to 1 part water, squeezing the solution through the pad. Replace a pad that remains brittle after soaking.
◆ Reinsert the pad into the holder and then slide the assembly back into the humidifier.
◆ Replace the top half of the cover, slide the plastic clips into place, and rehang the humidifier on the warm-air plenum. Press the humidifier down until it locks into place.
◆ Reconnect the water-supply line and overflow tube, then restore power to the furnace.

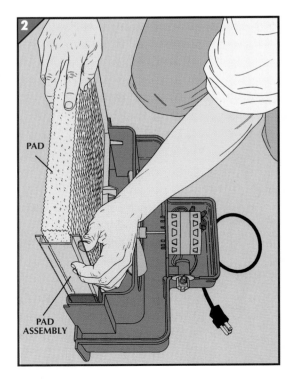

PAD

PAD
ASSEMBLY

# ADDING A HUMIDIFIER TO A FORCED-AIR SYSTEM

## ESTIMATING YOUR HUMIDITY NEEDS

| Thermostat setting | Weatherproofing of house | Gallons a day for each square foot |
|---|---|---|
| 65° | Tight | .003 |
| | Average | .004 |
| | Loose | .005 |
| 70° | Tight | .003 |
| | Average | .005 |
| | Loose | .006 |
| 75° | Tight | .004 |
| | Average | .007 |
| | Loose | .009 |

### Sizing a humidifier.

To determine the humidifier capacity you need, first evaluate the amount of insulation in your home's walls, floors, and ceilings and the presence—or absence—of cracks that admit outside air. A tight house has vapor barriers, close-fitting storm doors and windows, and new weather stripping and caulking. Then read across from your normal thermostat setting and multiply the right-hand figure by the total number of square feet on all the floors of your house including the basement.

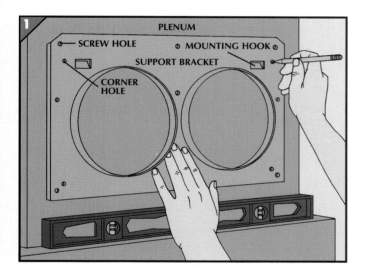

### 1. Positioning the stiffener.

◆ Shut off power to the furnace. Locate the humidifier support bracket, on the warm-air plenum. If your forced-air system includes an air conditioner, place the bracket on the side of the plenum parallel to the length of the A-frame coil and above the condensate pan.
◆ Level the bracket, then mark the plenum for all screw holes and for the four corner holes that define a rectangular cutout for the humidifier. Connect the corner marks.
◆ Drill holes for mounting screws, then make slits for tin snips by drilling a series of holes at each corner of the rectangle drawn earlier.
◆ Cut out the rectangle and screw the bracket to the plenum.

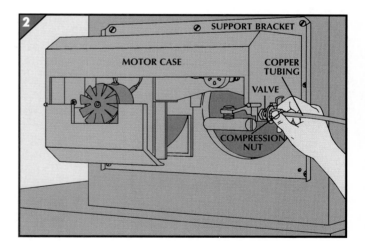

### 2. Mounting the humidifier.

◆ Separate the motor case of the humidifier from the rest of the cover. Fit the case onto the hooks on the support bracket and fasten the bottom of the case to the center of the bracket with the screw provided.
◆ To prepare the humidifier for the water connection, slip the nylon compression nut and sleeve provided with the unit over one end of a length of copper tubing.
◆ Push the end of the tube into the humidifier valve at the side of the case and tighten the compression nut with your fingers.

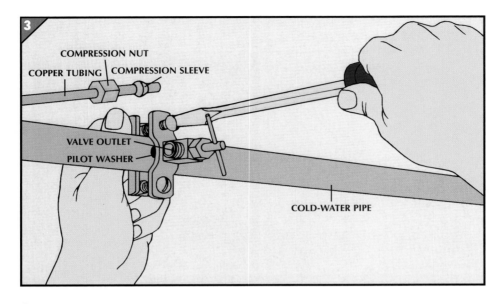

COMPRESSION NUT
COPPER TUBING   COMPRESSION SLEEVE
VALVE OUTLET
PILOT WASHER
COLD-WATER PIPE

### 3. Hooking up the water supply.
◆ Cut off the house water supply at the main shutoff valve and drain the system *(page 48)*.
◆ Find the copper or plastic cold-water pipe nearest the humidifier and tap into it with a self-tapping saddle valve *(above)*. To do so, remove one of the bolts on the saddle valve clamp, slip the rubber pilot washer over the pointed end of the valve stem, and slide the clamp onto the pipe. Replace the bolt in the clamp, tighten it, and turn the valve handle

clockwise until the stem pierces the pipe.
◆ Route the copper tubing to the valve and slip the brass compression nut and sleeve over the end. Insert the tubing's end into the saddle valve outlet and tighten the nut first with your fingers, then with a wrench.

If the nearest cold-water pipe is steel, use a plain saddle valve instead of the self-tapping variety. Bore a hole for the valve stem, denting the pipe beforehand with a center punch to keep the bit from slipping.

### 4. Mounting the humidistat.
◆ Position the mounting template that comes with the humidistat on the return-air duct at a convenient height. Drill the corners of the rectangle marked on the template, then cut out the sensor hole with tin snips.
◆ Carefully slide the sensor into the hole and push the humidistat case against the duct.
◆ Remove the cover of the case, drill holes for the mounting screws, and secure the humidistat to the duct *(right)*.

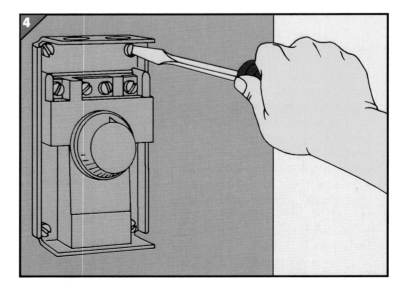

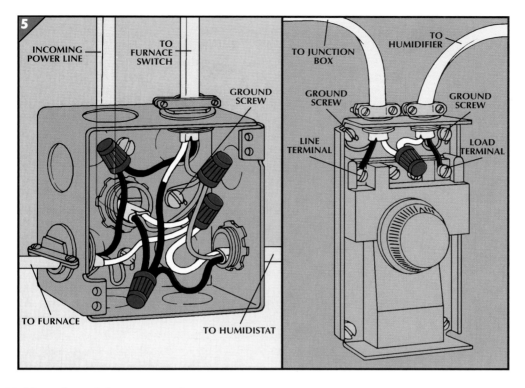

## 5. Making the wiring connections.

◆ To connect the humidistat and humidifier to the furnace junction box *(above, left),* first insert one end of a length of No. 14 cable, containing a ground wire, into the box, and the other end into the humidistat.

◆ At the box, use wire caps to attach the cable to the incoming power line—white wire to white, and black to black. With another wire cap, join the cable's and power line's copper ground wires to a bare copper jumper, then wrap the jumper around the box's ground screw.

◆ At the humidistat *(above, right),* wrap the cable's ground wire around a mounting screw and attach the black wire to the line terminal.

◆ Feed one end of a second cable into the humidistat, wrap the ground wire around another mounting screw, and attach the black wire to the load terminal. Join the ends of the two white wires with a wire cap.

◆ Feed the other end of the cable into the humidifier junction box and secure it with the connector washer provided. Attach the ground wire to the ground screw, then connect the black and white wires to the terminal board or the loose wires in the box.

## 6. Attaching the overflow drain.

◆ Slide the lower half of the humidifier cover up under the motor case. Attach it with the screws or clips provided.

◆ Fit one end of the drain tubing provided with the unit over the projecting sleeve of the overflow drain on the reservoir; secure it with a tubing clamp.

◆ Run the other end of the tubing to a nearby sink or basement floor drain and turn on the house water supply at the main valve.

◆ Restore power to the furnace, then set the humidistat to the relative humidity level desired and turn on the humidifier switch.

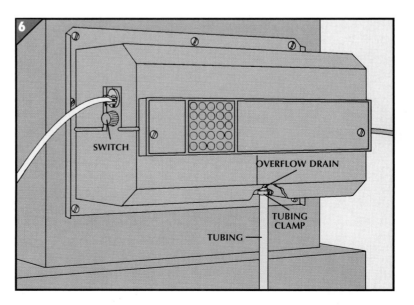

# Installing an Electronic Air Filter

While circulating air throughout a house's living spaces, a forced-air system also stirs up dust and pollen. An electronic filter inserted into the furnace's return-air duct can provide a much cleaner and more healthful indoor environment. The filter creates an electrical field in a self-contained "cell" that traps microscopic particles against metal plates.

These energy-efficient filters cycle on and off with the blower via a sail switch or by being wired to a junction box connected to the blower. When in use they consume about as much energy as a 40-watt light bulb.

**Selecting the Right Size:** Before purchasing an electronic air filter, remove the panel covering the furnace blower and measure the opening for the return duct. The filter opening may be up to 2 inches wider and 2 inches taller but may not be smaller.

Different brands vary little in operation (some may have two small cells rather than one large cell), but they may differ in thickness. Choose a model that requires the least modification to your existing ductwork.

**Keeping the Filter Clean:** An electronic filter should function silently—a crackling sound indicates that the cell needs cleaning. At a minimum, you should wash the filter elements twice a year. Turn the unit off, then remove the mesh pre-filter, which catches large particles, fibers, and pet hairs. Vacuum the mesh or wash it in hot water with a mild detergent.

Slide the cell out of the frame and soak it in hot, soapy water for 20 minutes before rinsing with a garden hose or immersing it in clean water for another 20 minutes. Let the cell drain dry.

Filter manufacturers suggest running the cells through a full dishwasher cycle, but some dishwasher manufacturers advise against doing so because the grit on the cell is abrasive and may damage a dishwasher.

 **TOOLS**

Tape measure
Screwdriver
Electric drill with
$\frac{1}{2}$" bit
Hammer
Tin snips
Cable ripper
Wire stripper

 **MATERIALS**

Self-tapping screws
($\frac{1}{2}$")
Wire caps
Sail switch kit
Silicone caulk

**SAFETY TIPS**

*Goggles will protect your eyes while you are drilling. To prevent cuts when working with ducts and cutting sheet metal, wear heavy work gloves.*

## 1. Measuring for the filter.

◆ After selecting an air filter of adequate capacity, measure the distance between the return-air duct coming down from the ceiling and the furnace wall on which the filter will be attached.

◆ Order from a sheet-metal shop a new return-air boot tailored to fit the return duct on one end and the air filter intake on the other, and also to accommodate the thickness of the device. If the space between the return duct and the furnace is too narrow to accommodate the filter frame, you must widen the space with a piece of ducting called a jump, as described opposite.

# How to Order New Ductwork

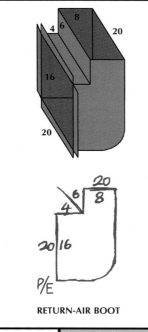

**RETURN-AIR BOOT**

**JUMP**

Sheet-metal shops and heating professionals have devised a notation system to express the size and shape of three-dimensional ductwork components as two-dimensional sketches. The key is knowing where to write the dimensions that are unseen in the simpler drawing.

The upper two illustrations at left show how the notation system applies to a return-air boot. The number 20, written twice outside the two-dimensional rendition of the shape, indicates front-to-back measurements—the third dimension not revealed in the drawing. The numbers 8 and 16 within the shape specify the other dimensions of the openings. Written on either side of a line that bisects the right angle, the numbers 4 and 6 stipulate the distance from the corner to the top and side of the boot. P/E is an abbreviation for plenum edge, a flange that is around one end of the boot that facilitates attaching it to the furnace or an electronic air filter.

The lower drawings show a jump, a section that replaces a straight piece of duct to provide added space between the return-air duct and the furnace for an air filter. As before, 20 refers to front-to-back measurements, and 8 refers to the width of the opening. Numbers between dotted lines outside the shape indicate the height of the jump (18 inches) and the additional space you need, in this case 6 inches.

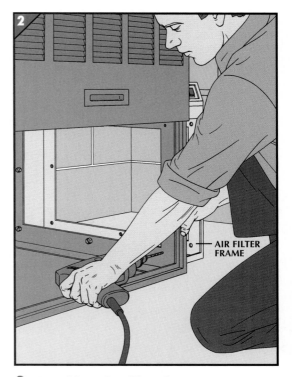

AIR FILTER FRAME

## 2. Attaching the filter frame.

◆ Detach the return-air boot from the return-air duct and the furnace.

◆ Remove the blower assembly from the furnace (page 37).

◆ Slide the electronic cell or cells and the metal mesh pre-filter out of the frame and set them aside.

◆ Position the frame next to the return-air opening in the side of the furnace.

◆ Working from within the furnace, drill pilot holes (above) for $\frac{1}{2}$-inch-long sheet-metal screws, or drive self-tapping screws through the furnace wall into each corner of the frame and then additional screws around the frame, 6 inches apart.

◆ Attach the new return-air boot—and jump, if you need one—to the existing return-air duct. Fasten the boot to the air filter frame with self-tapping screws.

## 3. Installing a sail switch.

◆ Affix the self-adhering template that comes with a sail switch *(photograph)* onto a straight, vertical segment of the return-air duct near the furnace. Make sure the airflow arrow points down.

◆ Drill mounting holes as indicated on the template, followed by overlapping holes at two corners of the switch opening as shown at right. Cut along the lines on the template with tin snips.

◆ Attach the sail to the switch mechanism, then squeeze the sail wires together and insert it into the cutout.

◆ Finally, attach the box to the duct with the screws provided.

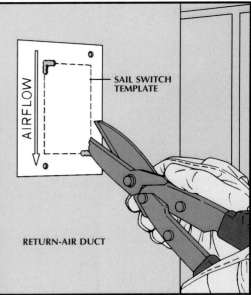

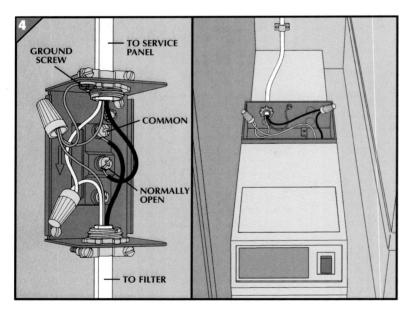

## 4. Wiring the switch and the filter.

◆ Remove the knockout tabs from the switch box's top and bottom.

◆ Fish cable from the service panel to the top of the switch. Run another length of cable from the bottom of the switch to the back of the filter. Strip the sheathing and insulation from the cable ends and clamp them to the switch box.

◆ At the switch *(above, left),* attach the black wire from the service panel to the terminal marked NO, for normally open, and the black wire from the filter to the terminal labeled COM, for common. Connect the two white cable wires with a wire cap. Join the two copper ground wires and a short jumper wire with a wire cap, then attach the jumper to the switch's ground screw.

◆ At the filter *(above, right),* join the cable and filter wires, black to black and white to white. Connect the ground wires and a short jumper with a wire cap, and secure the other end of the jumper to the filter's grounding screw.

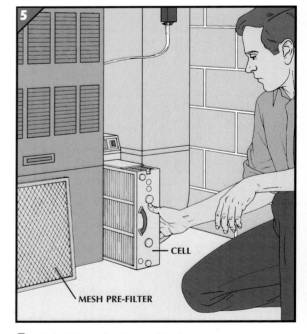

## 5. Finishing the installation.

◆ Align the cell on its track and slide it into the frame, then insert the metal mesh pre-filter into the groove next to the cell on the side closer to the return-air boot. Fit the air filter's door into place.

◆ Finally, seal the joints between the air filter frame and the furnace with a $\frac{1}{4}$-inch-thick bead of silicone caulk.

# ENERGY-SAVING DEVICES FOR VENTILATING A HOUSE

Today's tightly insulated homes are more energy efficient, but they also prevent fresh air from seeping into the house. Apart from the issue of breathing stale air, many homeowners are concerned under these circumstances about the buildup of radon, a naturally radioactive carcinogen. Heat-recovery ventilators, or HRVs, address these issues by providing a supply of fresh air without significantly increasing winter heating costs. HRVs have become standard features in many new homes in Canada and the northern United States.

**How HRVs Work:** In a typical installation *(below)*, the HRV is located near the forced-air furnace and wired to a switch in the living space. Branch ducts connect the HRV to the furnace's main return duct, and to supply and exhaust vents on an outside wall.

Stale air from the return-air duct *(red arrows)* is diverted to the HRV and vented outdoors. Meanwhile, fresh replacement air from outside the house *(blue arrows)* travels through the HRV en route to the return-air duct. Inside the HRV, the two air streams pass independently through a heat exchanger, which extracts heat from the stale air and transfers it to the fresh. Some HRVs can recover more than 80 percent of the heat in the stale air before expelling it.

**Rejecting Humidity:** Another type of ventilator, suitable for moderate and warm climates, has the additional capability of transferring moisture between air streams. In winter, this unit works in the same way an HRV does; in summer, it removes humidity from outdoor air before routing it through the central air conditioner. These versatile HRVs are sometimes called energy-recovery ventilators, or ERVs.

**Buying Tips:** An HRV can be added to a forced-air furnace or installed as an independently ducted system in homes with other types of heat. In either case, the HRV should be installed by a licensed contractor trained to integrate the ventilator with the existing heating and cooling system.

The heat-transfer efficiency ratings of most HRVs and ERVs sold in Canada and the United States are available from distributors or heating-and-cooling contractors.

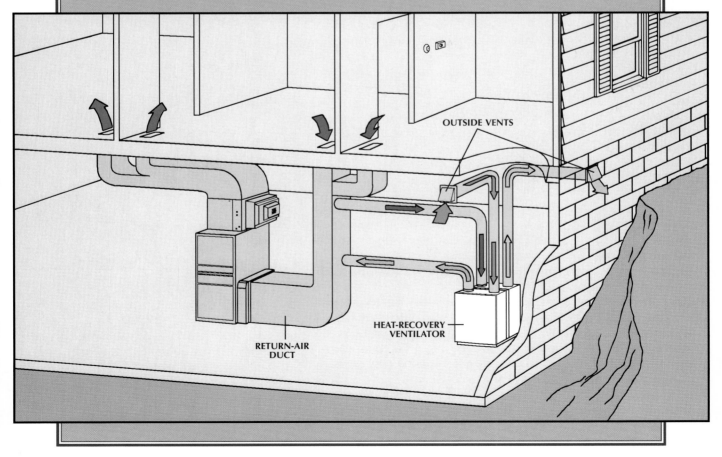

OUTSIDE VENTS

HEAT-RECOVERY VENTILATOR

RETURN-AIR DUCT

**A** forced hot-water heating system is built for reliability, and regular servicing should keep it running smoothly for years. If problems do arise, the system may have to be drained to make repairs, as explained below. Draining may also be necessary if you want to add valves to improve the system *(page 51)*.

Components of a hot-water system rarely fail, but when breakdown occurs, it usually involves the pump motor, the coupler holding the motor to the pump shaft, or the pump seal. Remove the burned-out or broken part *(pages 49-50)*, and take it to a heating-supply dealer for an identical replacement.

**Maintenance Basics:** Once a year, before starting the boiler, bleed the convectors to rid the system of air *(page 11)*. If your circulator pump is not self-lubricating *(box, page 50)*, put a few drops of No. 20 nondetergent electric-motor oil in the oil cups at both ends of the motor and on the top of the bearing assembly between the motor and pump body.

**Checking Water Pressure:** During the heating season, periodically examine the combination gauge on the side or front of the boiler. Depending on the size of your house, the pressure can safely range from as little as 3 pounds per square inch when the water cools and contracts, to about 30 pounds when it heats and expands. The expansion tank *(page 51)* provides a cushion of air for the expanding and contracting water. A conventional tank has a top layer of air in direct contact with a layer of water, while a diaphragm tank keeps the air layer at the bottom, separated from the water by a rubber membrane.

If the movable "pressure" pointer on the gauge drops below the stationary "altitude" pointer, increase the pressure in the system by adding water to a conventional tank, or recharging the air in a diaphragm tank. If the movable pointer passes 30 pounds, there is too much pressure. Call for service if you have a diaphragm tank. Recharge a conventional tank with air *(page 51)*.

⚠️ **CAUTION** *Old boilers may have asbestos insulation in the liner or around the pipes. To test, mist a small area with a solution of 1 teaspoon of low-sudsing detergent per quart of water, then remove a small sample and take it to a local lab certified by the National Institute of Standards and Technology. If the test is positive, hire a plumber or other professional licensed to handle asbestos.*

 **TOOLS**

| | |
|---|---|
| Hex wrench | Bicycle pump |
| Open-end wrench | Hacksaw or tube |
| Box or socket | cutter |
| wrench | Propane torch |
| Tire gauge | Flameproof pad |

 **MATERIALS**

| | |
|---|---|
| No. 20 nondetergent electric-motor oil | Solder |
| | Flux |
| | Three nipples |
| Garden hose | Shutoff valve |
| Wood block | Union |
| Bucket | |

🪖 **SAFETY TIPS**

*When soldering, wear gloves and eye protection.*

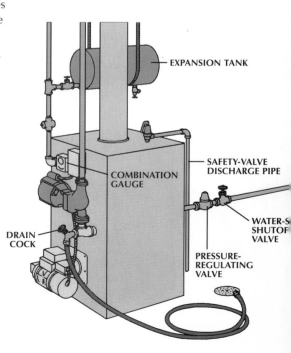

EXPANSION TANK

SAFETY-VALVE DISCHARGE PIPE

COMBINATION GAUGE

DRAIN COCK

WATER-S SHUTOF VALVE

PRESSURE-REGULATING VALVE

## Draining and refilling the system.

◆ Turn off the power to the boiler at the master switch and the service panel.

◆ When the combination gauge indicates the water in the system is lukewarm, close the water-supply shutoff valve. Attach one end of a garden hose to the boiler drain cock, and run the other end of the hose to a floor drain.

◆ Open the boiler drain cock and the bleed valves of all the convectors on the upper floors of the house and let the water drain.

◆ Refill the system by closing the convector bleed valves and the boiler drain cock and opening the water-supply valve. If there is a pressure-regulating valve on the line, the flow will stop automatically when the system is full. Otherwise, fill until the combination gauge's movable pointer corresponds to the position of the stationary pointer.

◆ Bleed all of the heating units on the upper floor. If you do not have a pressure regulator, have someone bleed each unit while you watch the gauge.

## Troubleshooting Guide

| PROBLEM | REMEDY |
|---|---|
| No heat. | Replace fuse or reset circuit breaker.<br>Bleed air from convectors *(page 11)*.<br>Replace coupler *(page 50)*.<br>Replace circulator pump motor *(below)*. |
| No heat, motor housing hot, burning odor. | Replace motor *(below)*. |
| Heat uneven throughout house. | Bleed air from convectors *(page 11)*.<br>Replace coupler *(page 50)*.<br>Replace circulator pump motor *(below)*. |
| Not enough heat, convector is lukewarm. | Vacuum and straighten convector fins *(page 11)*.<br>Bleed air from convectors *(page 11)*.<br>Replace circulator pump motor *(below)*. |
| Circulator motor noisy. | Lubricate circulator pump and motor. |
| Circulator motor sounds like a chain being dragged through system. | Replace coupler *(page 50)*. |
| Water spills from safety-valve discharge pipe. | Recharge conventional expansion tank *(page 51)*.<br>Replace diaphragm-type expansion tank. |
| Circulator pump leaks. | Replace seal *(page 50)*. |

# REPLACING PUMP PARTS

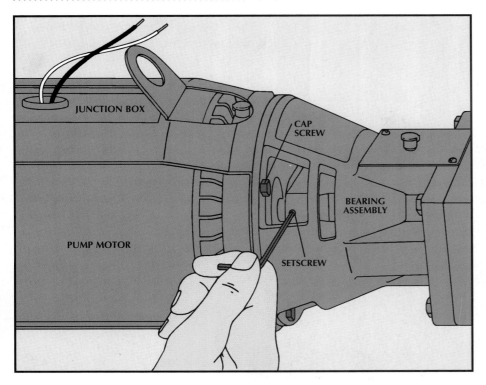

### Installing a new motor.

◆ Turn off the power to the boiler at the master switch and service panel.
◆ Remove the junction box cover from the pump motor, and disconnect the wires.
◆ With a hex wrench, remove the setscrew holding the coupler to the motor shaft.
◆ Grip the motor in one hand, and using an open-end wrench, loosen the cap screws that hold the motor to the bearing assembly.
◆ Back the motor out, leaving the coupler attached to the pump shaft.
◆ Fit the free end of the coupler onto the shaft of the new motor.
◆ Holding the new motor against the bearing assembly, reinsert the cap screws and secure the coupler setscrew.
◆ Reconnect the junction box wires and replace the junction box cover.

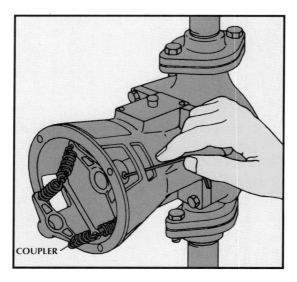

## Changing the coupler.

◆ Remove the motor *(page 49)*.
◆ With a hex wrench, loosen the setscrew that is holding the coupler to the pump shaft, and slide off the coupler.
◆ Secure one end of the new coupler to the pump shaft with the setscrew.
◆ Replace the motor and attach the other end of the coupler to the motor shaft.

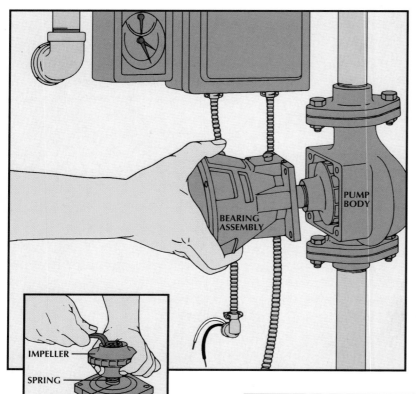

## Replacing the pump seal.

◆ Drain the system *(page 48)*. Alternatively, turn off the power to the boiler at the master switch and service panel, then cut off the water to the pump by shutting the valves above and below it.
◆ Remove the motor and the coupler *(page 49 and above)*.
◆ Undo the cap screws holding the bearing assembly to the pump body, and pull the assembly out.
◆ With the bearing assembly standing on a wood block for support *(inset)*, turn a box or socket wrench clockwise to loosen the nut holding the impeller to the pump shaft.
◆ Slide off the impeller and spring, and save them.
◆ Pull off the brass seal. Slide a new one onto the shaft, and press it tight.
◆ Attach the old spring and impeller with the nut and washer, and reassemble the pump.
◆ Refill the system, or open the valves at the pump. Restore the power.

### A LOW-MAINTENANCE PUMP
..............................................
Many modern circulator pumps are self-lubricating (that is, the motor and bearings require no oil), and they also may have fewer parts than older models, reducing the chances of a breakdown. Some, like the pump at left, also offer a choice of speeds, allowing you to adjust the rate at which hot water circulates. By altering the pump's speed to closely match your system's needs, you can reduce water noise in the pipes and at the same time save energy.

# SERVICING THE EXPANSION TANK

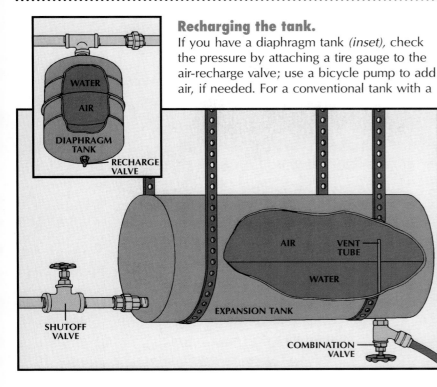

**Recharging the tank.**
If you have a diaphragm tank *(inset),* check the pressure by attaching a tire gauge to the air-recharge valve; use a bicycle pump to add air, if needed. For a conventional tank with a shutoff valve and a combination valve, turn off power to the boiler at the master switch and service panel, close the shutoff valve on the line between the tank and boiler, run a hose from the combination valve to a nearby drain, open the valve, and empty the tank. For a conventional tank that lacks a combination valve, open the plug or drain cock at the tank's base and let the water empty into buckets. For one lacking a shutoff valve, drain and refill the entire system *(page 48).*

You can add a shutoff valve as described below. You can also add a combination valve:
◆ Turn off the power to the boiler at the master switch and service panel, empty the tank, and close the shutoff valve.
◆ Cut the vent tube of the valve to two-thirds the height of the expansion tank.
◆ Remove the plug or drain cock from the tank's base, and screw the combination valve into the opening.

# ADDING A SHUTOFF VALVE

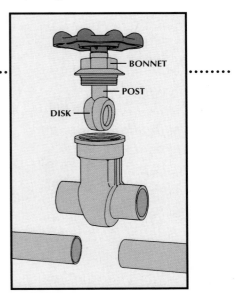

**A fitting of copper.**
◆ Drain the system *(page 48).*
◆ Working on the line between the expansion tank and the boiler, use a tube cutter or hacksaw to cut out a section 1 inch shorter than the length of the shutoff valve.

◆ Unscrew the bonnet from the valve body with a wrench and lift out the disk assembly; otherwise the soldering heat may warp the disk or post.
◆ Solder the valve to the pipe and, once it has cooled, replace the disk assembly.

**A threaded steel fitting.**
◆ Drain the system *(page 48).*
◆ Cut the line near the inlet of the expansion tank, and unscrew the pieces of pipe from their fittings at both ends of the section.
◆ Attach a 2-inch nipple—a short pipe with threads at both ends—to a shutoff valve.
◆ Screw an assembled union to the other end of the nipple, undo the ring nut, and lift off the free union nut.
◆ Attach a 6-inch nipple to

the expansion tank. Add the valve assembly by slipping the ring nut over the nipple and attaching the remaining union nut.
◆ Slide the ring nut over the union nuts and tighten it *(right).*
◆ Measure from the free end of the shutoff valve to the closest fitting, and buy a nipple to fit the gap. If it is not a standard length, have a metal shop cut and thread a pipe.
◆ Refill the system *(page 48).*

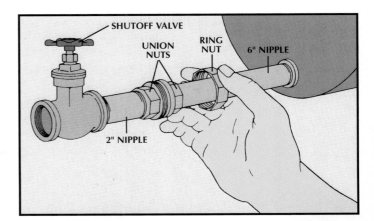

# New Heat Where You Need It

When you alter your floor plan by building an addition, enclosing a porch, or finishing a basement or attic, you need to decide how to heat the new space. The choices are extending your existing forced-air or hot-water system or installing one or more independent space heaters. Depending on the climate and orientation of your home, you might also consider taking advantage of solar energy.

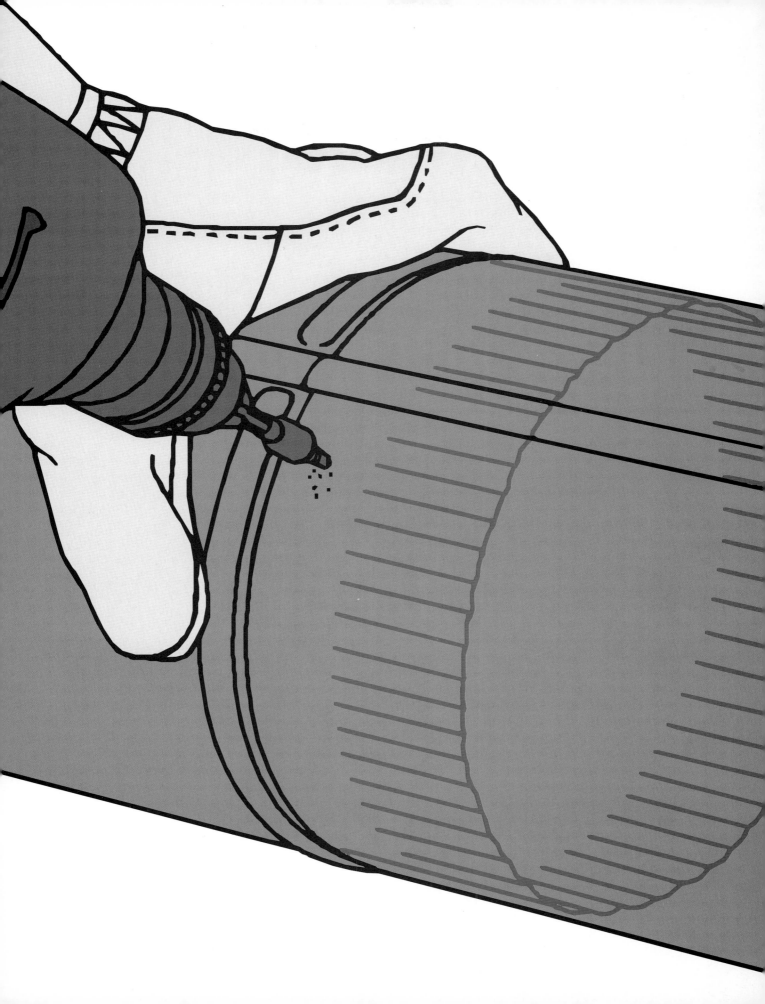

Some ducts in a forced-air system carry warm or cool air to the house. Others are returns, taking air back to be heated or cooled. If your house has this type of system, in some cases you need only add a register—a new opening in a duct—to make a house more comfortable in all weathers.

**Mapping Your Ductwork:** Before you add a register, you must find out where all the ducts are and what they do. Fire up the furnace. While it is running, touch the ducts connected directly to it; supply ducts will feel warm, returns cold. Exposed ducts in the basement or attic can be identified in the same way. Trace the routes of the ducts, noting where they enter rooms.

Next, go to the registers. Remove them and examine the spaces behind. Each register connects to a duct either directly *(pages 54-56)*, or by means of an intervening extension box *(page 57)*, or boot. Usually you can see whether the duct runs down to the register, up to it, or horizontally. Identify the ones that blow air in and the ones that suck air out.

With the map of your system completed, you can choose the points to tap. If it is possible, place new registers near sources of winter cold—exterior walls, windows, and doors. An average-size house needs one return register per floor, but do not add one in a bathroom or kitchen, since moisture and odors could be spread through the house.

**TOOLS**

Scissors
Electric drill
Utility knife
Screwdriver

Hammer
Tin snips
Broad-billed pliers
Keyhole saw or
saber saw

**MATERIALS**

Cardboard
Stiff wire
Extension box
or register collar

Wall or floor
register
Self-tapping sheet-
metal screws

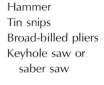

**SAFETY TIPS**

*When working with sheet metal, protect your hands with heavy-duty work gloves and wear a long-sleeved shirt.*

# A DIRECT CONNECTION TO A WALL DUCT

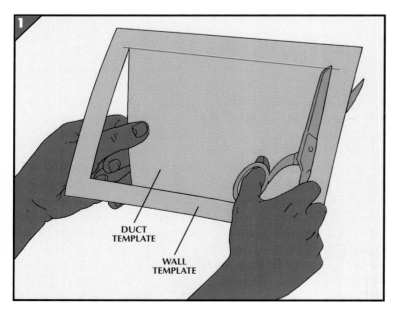

DUCT
TEMPLATE

WALL
TEMPLATE

**1. Templates for the wall and duct.**
◆ Outline the register's front on a piece of stiff cardboard and cut out the outlined rectangle.
◆ When adding a supply register, measure the collar on the back of the register and add $\frac{1}{2}$ inch to both its length and width. For a return register, add $\frac{1}{2}$ inch to the length and width of the louvered area.
◆ Transfer these dimensions to the center of the cardboard rectangle, and cut out the inner rectangle *(left)*, forming templates for marking wall and duct openings.

DUCT TEMPLATE

WALL TEMPLATE

## 2. Completing the templates.

On the duct template, mark lines $1\frac{5}{8}$ inches in from the edges, and snip out the corners.

## 3. Cutting a hole.

◆ Tape the wall template to the surface concealing the duct.

◆ Drill small holes through the wall inside the template and probe with a stiff wire to check that the inner rectangle does not extend beyond the duct *(right)*. If necessary, move the template to fit completely over the duct.

◆ Mark the inner shape of the template on the wall surface and cut around the outline with a utility knife.

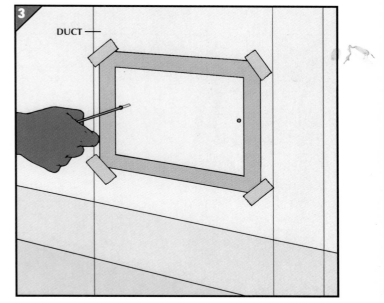

DUCT

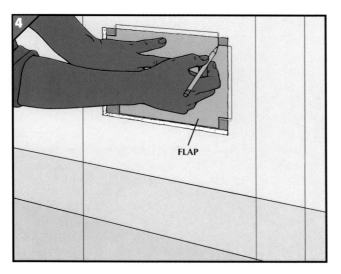

FLAP

## 4. Marking the duct.

◆ Set the duct template against the duct inside the wall opening.

◆ Mark the cutout corners of the template on the duct, then fold back the flaps to make a rectangle.

◆ Set the rectangle against the duct with its corners touching the points you have marked, and draw its outline on the duct.

◆ Draw diagonal lines from the corners of the rectangle to the corners of the wall opening.

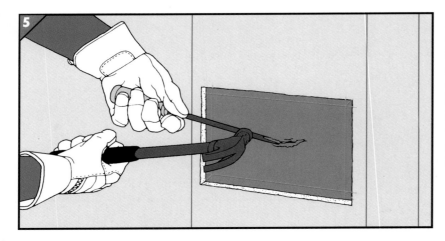

### 5. Starting a hole.

◆ Press the edge of a screwdriver blade against the duct near the center of the rectangle you have drawn.
◆ Hit the screwdriver shank with a hammer to make a hole large enough for the tips of a pair of tin snips.
◆ Cut around the rectangle, then snip along the diagonal lines at the corners to create four flanges.

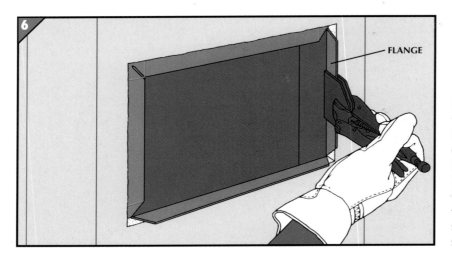

FLANGE

### 6. Making flanges.

◆ With broad-billed pliers, bend the flanges around the edges of the wall opening to lie flat against the wall. Finish the job with a hammer, if necessary.
◆ Attach the register with self-tapping sheet-metal screws *(page 60).*

## A Range of Registers

The supply registers at right typify the models generally used to expand a forced-air system. All have louvers to direct the flow of air into a room. Rectangular supply registers for walls and floors have a control to regulate air flow. Ceiling registers, whether they are circular or rectangular, as shown here, generally do not. Furthermore, they have wider louvers than wall and floor registers; the larger openings create air turbulence to prevent heated air from collecting at the ceiling.

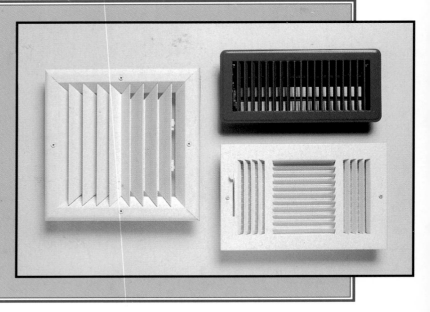

# AN EXTENSION BOX FOR A RECESSED FLOOR DUCT

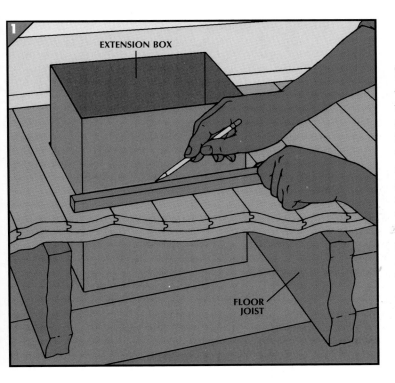

**EXTENSION BOX**

**FLOOR JOIST**

## 1. Cutting the box to fit.

◆ Have a sheet-metal shop prepare a box that fits the collar of the register and that is at least 6 inches longer than the height of the floor joists.

◆ Prepare templates for the floor and duct openings by the methods shown on pages 54 and 55, Steps 1-3, then saw a hole in the floor with a keyhole or saber saw.

◆ Stand the box on the duct and mark lines 1 inch above floor level on all four sides; a scrap of 1-inch molding makes a handy guide (left).

◆ Trim the box at the marks you have made, then snip the corners and bend the sides to form 1-inch flanges.

**BOX FLANGES**

## 2. Installing the register.

◆ Mark the duct using the template (page 55, Step 4), then cut an opening as shown in Step 5 (opposite).

◆ Make diagonal cuts $1\frac{5}{8}$ inches long at the corners of the duct opening, and bend the edges upward to make flanges perpendicular to the duct (Step 6, opposite).

◆ Set the box on the duct, flanges down (left), with the duct flanges inside the box.

◆ Reach inside the box and fasten the duct flanges to the box on two opposite sides with self-tapping screws (page 60). If space is limited, punch holes in the flanges and attach the screws with a short-handled screwdriver.

◆ Set the register into the floor opening.

# Extending a Forced-Air System

Adding one or two duct runs to an existing system is a matter of basic carpentry and sheet-metal work. All of the parts of a duct run are standardized and easy to hook up. And the carpentry is simplified when you run round ducts outside the wall instead of fitting shallow rectangular ducts between wall studs.

**Where to Begin:** The starting point for a new run will depend on the design of your existing air-distribution system. In a radial system, all duct runs—including new ones—start at the furnace plenum and radiate out-ward toward the rooms of the house. In an extended-plenum system, one or more large rectangular ducts extend from the plenum, and all ducts for individual rooms branch from these main ducts.

**Selecting Duct:** Round duct, called pipe, should be at least 6 inches wide. Duct made from rigid metal provides the quietest and most efficient air flow and is the best choice for long runs. You can use flexible plastic or aluminum duct to negotiate wide turns or odd angles, or for short runs in hard-to-reach areas.

With a combination, you can extend a duct to any point in a house.

**Concealment Techniques:** Whenever possible, hide new duct by running it inside a closet or above a suspended ceiling. More often, you will have to build an enclosure to hide it, a task that is minimized by ending each duct run at a floor register. Although floor registers work best under windows, you may decide to end runs at interior walls if the carpentry required to reach an exterior wall makes such a location impractical.

 **TOOLS**

Tin snips
Electric drill with
   12" bit extension
   and magnetic
   nutdriver ($\frac{5}{16}$")

Plumb bob
Wallboard saw
Carpenter's square
Saber saw
Broad-billed pliers
Minihacksaw

 **MATERIALS**

Duct (6" round and
   flexible)
Starter collar
Sheet-metal screws
Hangers

Joist-hanger nails
Register boot
Floor register
Lumber (2 x 2, 2 x 4)
Wallboard
Joint compound and
   tape

 **SAFETY TIPS**

*Wear work gloves and a long-sleeved shirt when handling duct. Use goggles when cutting and patching ceiling openings.*

## How ducts run.
The drawing at right shows how new duct runs can be added to either a radial *(left)* or an extended-plenum *(right)* system. In either case, the new runs begin with a fitting called a collar. A takeoff collar is generally needed for the top of an extended plenum; its adjustable elbow can start a run of duct in almost any direction. A run may go directly to a single room or branch with a T or Y fitting to serve more than one room. Each run or branch is fitted with a damper to control flow. For turns, elbows are available preset at 45- and 90-degree angles or in flexible designs that can be turned to the desired angle. Turns can also be made with flexible duct, like the length in the run at right. Each run ends in a boot for a floor register.

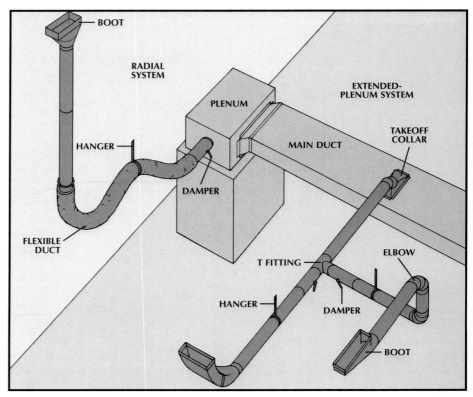

# STARTING A RUN

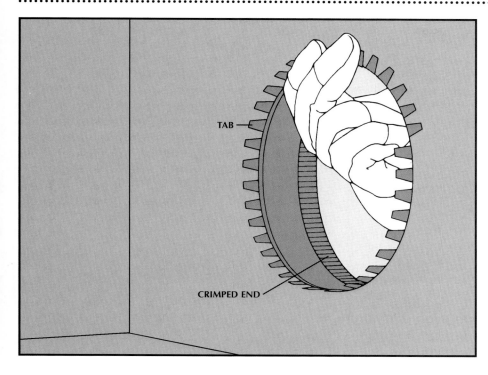

TAB

CRIMPED END

### A straight collar.
◆ At least 1 inch below the top of the furnace plenum, cut a hole exactly the size of the collar (page 56, top).
◆ Slide the collar into the hole, tabbed end first, until the projecting bead around the exterior of the collar fits tightly against the side of the plenum.
◆ Reach inside the collar and fold the tabs to lie flat against the inside of the plenum. Connect the plain end of the first duct section in the run to the crimped outer end of the collar.

### A takeoff collar.
◆ Choose a place for the collar on the top or the side of the extended plenum, making sure that the collar elbow clears the joists or ceiling above by at least 1 inch.
◆ Cut a rectangular hole just big enough to admit the collar's fold-over tabs, then slide the takeoff into the hole until the flange rests on the plenum.
◆ Reach into the circular end of the collar and fold the tabs flat against the inside of the plenum.
◆ On the outside of the plenum, drill holes through two opposite sides of the flange and the plenum beneath it, and secure the collar to the plenum with sheet-metal screws.

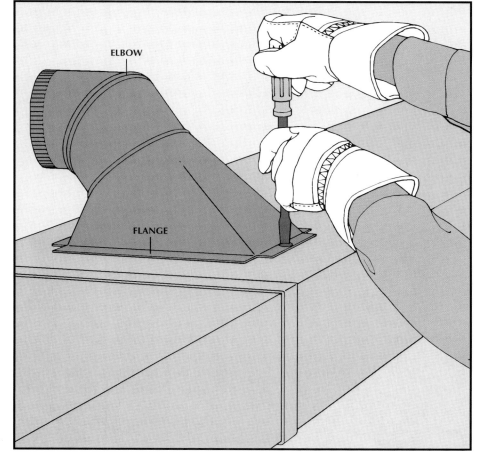

ELBOW

FLANGE

# ASSEMBLING DUCT SECTIONS

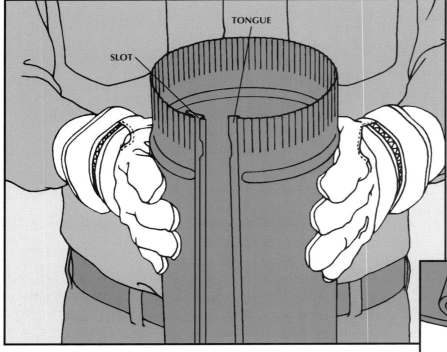

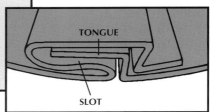

## A snap-lock assembly.
◆ Round duct is generally shipped and sold in open sections. Before closing the seams, measure the entire duct run and shorten one section, if necessary, by snipping a length off the plain end. If the snips pinch the seams flat, pry them open with a screwdriver.
◆ Starting at one end of the section, shape the duct into a cylinder. Then push the tongue into the slot until the seam snaps shut *(inset)*.
◆ Repeat at 1- or 2-foot intervals.

## A hammer-lock assembly.
◆ When the seam consists of two U-shaped edges, as shown below, shape the entire duct section with your hands and hook the edges together lightly.
◆ Hang the duct from a 2-by-4 supported on edge by sawhorses. Starting at either end, hammer the seam shut. Hammer blows can easily unfasten the still-unsealed parts of the seam. Check constantly for separation as you hammer along the section; an error may force you to reopen an entire seam.

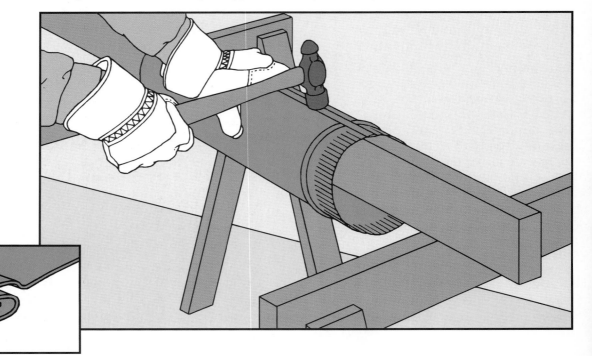

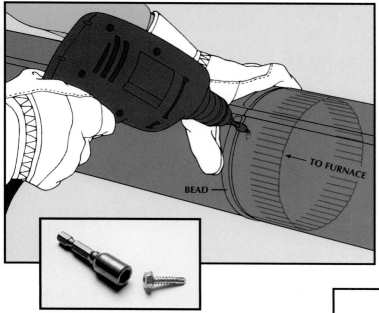

## Joining full-length duct sections.

Make sure that crimped ends point away from the furnace. Doing so ensures an airtight seal and eliminates vibration.

◆ Slide the crimped end of one duct into the plain end of the next section, aligning the seams of the two sections. Push until the bead of the crimped section touches the edge of the plain one.

◆ Just outside the bead, drive two No. 10 self-tapping screws, $\frac{1}{2}$ inch long, through opposite sides of the joint. Use a $\frac{5}{16}$-inch magnetic nutdriver (photograph) in an electric drill to hold and drive the screws.

◆ Wrap several turns of duct tape around the bead to seal the joint.

## A drawband for the final section.

To make the last joint in a duct run, use a drawband—a flexible steel collar tightened by nuts and bolts.

◆ Trim the plain end from a section of duct to make a piece about 2 inches longer than the gap it is to fill. The plain end should meet or nearly meet, but not overlap, the crimped end of the adjoining duct or fitting.

◆ Connect the crimped end of the shortened section, as shown at right.

◆ Slip the drawband all the way onto the plain end, align the duct sections, then slide the collar onto the crimped end until it touches the bead.

◆ Tighten the bolts, using a screwdriver and a wrench as needed, then seal the connection with duct tape.

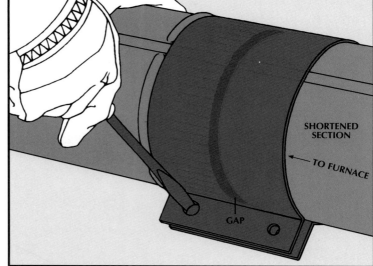

# DAMPERS AND HANGERS

## Installing a damper.

Each duct run should have a damper, preferably near the furnace, to shut off heat or air conditioning to a room and to balance the system (pages 8-9). You can buy 2-foot duct sections with factory-installed dampers, or you can buy dampers separately and install them yourself as shown here.

◆ Drill or punch holes for spring-loaded damper shafts in opposite sides of the duct, recessed at least 6 inches from the plain end.

◆ Retract the shafts in their slots, set the damper in the duct, and release the shafts.

◆ Align the damper so that the shafts spring through the holes you have made, then attach the damper handle to the more accessible shaft.

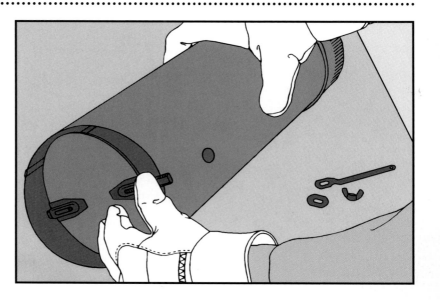

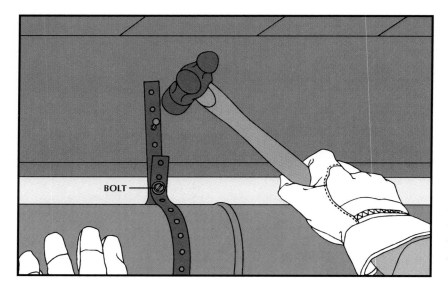

## Installing a hanger.

To support horizontal duct runs from above, use flexible metal straps called hangers. Available in lengths to fit ducts of any diameter, hangers are perforated to accept nails and bolts.

◆ Loop one end of each hanger snugly around the duct, thread a bolt through opposite perforations near the top of the duct, and fasten with a nut.

◆ Fasten the other end of the hanger to the supporting surface with a joist-hanger nail, keeping a clearance of at least 1 inch between the duct and the joist or ceiling.

◆ Install hangers no more than 10 feet apart; if the duct sections are relatively short—6 feet or less—shorten the intervals accordingly.

BOLT

# TECHNIQUES FOR A NEW DUCT RUN

### Anatomy of a concealed duct.

Air from a basement furnace passes to the second floor through a round duct in the corner of a first-floor room. Upstairs, air enters the room through a boot, a metal transition fitting from the 6-inch round duct to a rectangular 4- by 10-inch floor register. Because the boot is wider at its base, it is installed from below to avoid having to cut an oversized opening in the second-story flooring. Nailers installed in the ceiling opening provide surfaces for fastening a ceiling patch, required by building codes even if the hole will be concealed. The duct is boxed in by two adjoining walls and by a wallboard-covered frame of 2-by-2s and 2-by-4s fastened to the adjacent walls.

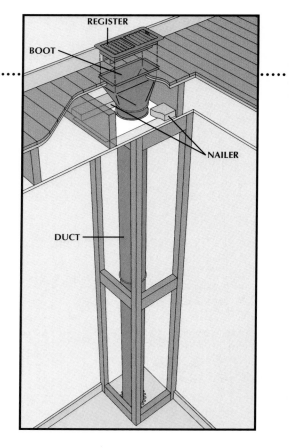

REGISTER

BOOT

NAILER

DUCT

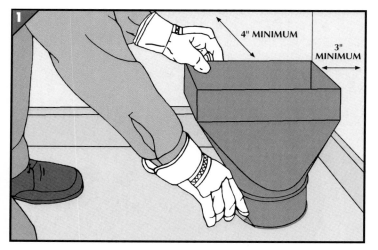

4" MINIMUM

3" MINIMUM

### 1. Positioning.

◆ Set the register boot on the floor in the first-floor corner selected for enclosing the new duct. Place the boot with its long side at least 4 inches from the longer wall of the room (to optimize air flow) and its short side at least 3 inches from the adjoining wall.

◆ Outline the base of the boot on the floor (left). At the center of this outline, drive a nail through the floor.

◆ In the basement, check that there are no ducts, pipes, joists, or other obstructions within $3\frac{1}{2}$ inches of the nail. If necessary, move the circle.

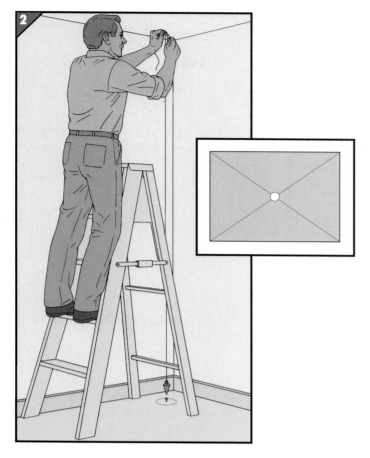

## 2. Cutting a ceiling opening.

◆ Using a plumb line, mark the ceiling directly above the nail in the floor.

◆ Drill a $\frac{1}{4}$-inch hole at the mark, and with a length of stiff wire, check for obstructions within $3\frac{1}{2}$ inches of the hole.

◆ Enlarge the ceiling hole to a diameter of $\frac{3}{4}$ inch, and with a 12-inch bit extension, drill a matching hole through the flooring overhead.

◆ Make a cardboard template *(inset)* 14 inches long and 10 inches wide. Draw the diagonals of the rectangle and cut a quarter-size hole where they cross in the center of the template.

◆ Center the template over the ceiling hole and parallel to both walls, then outline the template on the ceiling.

◆ Cut the ceiling along the marks.

## 3. Making the register opening.

◆ On the second floor, position a carpenter's square with one arm against the wall that parallels the length of the boot and one edge of the other passing through the center of the hole drilled in Step 2.

◆ To accommodate the offset built into the boot, mark the floor $\frac{5}{8}$ inch from the edge of the hole nearest the wall *(right)*.

◆ Make a 4- by 10-inch cardboard template. Center it over the mark on the floor, aligning it with the wall, then outline it on the floor.

◆ Drill holes at the outline's corners and cut out the opening with a saber saw. Use a metal-cutting blade to saw through any flooring nails in the way.

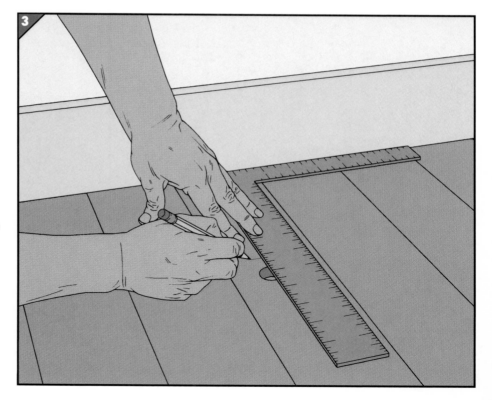

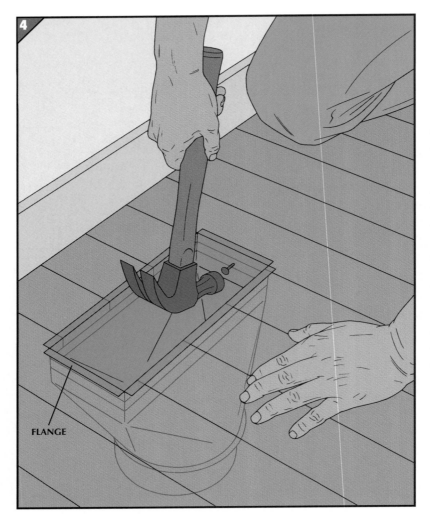

FLANGE

## 4. Installing the register boot.
◆ With broad-billed pliers *(page 56, Step 6)*, bend flanges $\frac{1}{2}$ inch wide around the boot's register opening, then bend them back to a position that is almost vertical.
◆ Have a helper on the first floor maneuver the boot into the second-floor opening, then bend the flanges to lie against the floor.
◆ Secure the boot by driving a 1-inch roofing nail through each end of the boot and into the subflooring.
◆ On the first floor, draw a new circle $\frac{1}{2}$ inch wider than the original boot outline. Drill a starter hole and cut along the outline with a saber saw.

## 5. Making a ceiling patch.
◆ If necessary, widen or lengthen the ceiling opening to the midpoints of the closest joists.
◆ Along the other sides of the opening, install 2-by-4 nailers between the joists. Toenail the 2-by-4s facedown so that they extend half their width into the opening.
◆ Using a straightedge and pencil, extend all four sides of the opening several inches across the ceiling.
◆ In an oversize wallboard patch, cut a 6-inch circular hole at the approximate location of the boot's round end.
◆ Hold the patch against the ceiling, hole aligned with boot, and mark the edges of the patch where they cross the lines drawn on the ceiling *(right)*.
◆ Join the edge marks with straight lines, then cut along them to complete the patch. Set the patch aside.

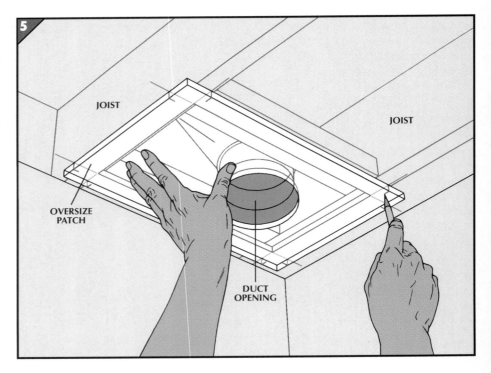

JOIST

JOIST

OVERSIZE PATCH

DUCT OPENING

## 6. Installing the duct and patch.

◆ Cut the rivets at the seam of a starter collar with a mini-hacksaw, fold the tabs outward to a right angle, and fit the collar into the floor hole, crimped edge down.

◆ Slide the plain end of a full-length (5-foot) section of duct through the collar so that 5 or 6 inches protrude into the room below. Hold the collar tightly around the duct and temporarily nail a few tabs to the floor.

◆ Measure from the bead at the duct's upper end to the bead on the register boot *(right)*. Cut a section of duct that length.

◆ Nail the patch in place. Pull the nails in the tabs to loosen the collar; have your helper lower the 5-foot section 6 inches.

◆ Fit the crimped end of the cut section through the patch and into the boot. While your helper pushes the top of the 5-foot section into the bottom of the cut section, retighten the collar and nail every third tab to the floor.

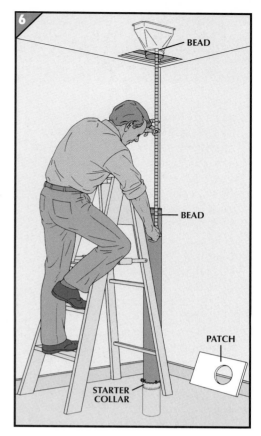

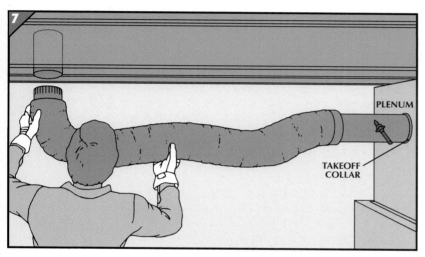

## 7. Connecting to the plenum.

◆ Cut a hole in the sheet-metal plenum of the furnace and install a starter collar in the opening *(page 59)*.

◆ Join the plain end of a 2-foot duct section containing a damper to the crimped end of the collar.

◆ Connect the plain end of some flexible duct to the other end of the damper section, then insert the crimped end of the flexible duct into the plain end of the rigid duct protruding into the basement. (If the basement is unheated, use flexible duct with factory-installed insulation and vapor barrier.)

◆ Seal all joints with duct tape and support the damper section and the flexible duct with duct hangers.

## 8. Enclosing the duct.

◆ Measure and cut three 2-by-2 uprights and 2-by-4 blocking for a frame big enough to conceal the ceiling patch. Butt-nail blocking between the uprights at floor and ceiling level and midway between. Make sure the uprights are vertical and the braces horizontal.

◆ Remove baseboard from the corner, and nail the uprights to wall studs with $3\frac{1}{2}$-inch nails, or drill holes for wallboard anchors. If necessary, shim the frame to make it plumb.

◆ Cut wallboard to fit the sides of the frame and fasten it with wallboard screws. Cover screwheads with joint compound, and finish seams with tape and joint compound. Install baseboard around the enclosure and paint it.

Baseboard units, screwed to wall studs, give off radiant heat to warm a room. Heaters that tuck into the space beneath a cabinet and those that are recessed into walls have built-in fans to force air over the heating element and into the room. A heater or heat lamp can be mounted on a ceiling to provide instant heat to the area directly beneath it. Many of these are available in 120- and 240-volt models. Talk with a dealer, describing the space you want heated, and how often and for how long you expect to use the heater.

**Wiring the Heater:** Some smaller baseboard heaters plug directly into a wall receptacle. But if you want a larger unit—or need more than one—run cable from the service panel and install the more efficient 240-volt model as shown on pages 71 and 72. Connections at the service panel are best left to a professional unless you have had extensive experience in electrical work. However, the remainder of the installation and wiring involves only elementary techniques; doing it yourself may save considerable expense.

The capacity of the heater will determine the size of the cable for the new circuit. Check with the heater supplier for the correct size cable to use.

**Controlling the Heat Output:** All permanently installed heaters need either thermostats or switches. Many come with such controls already built in, but others require separate controls, such as a timer switch for a ceiling heater, a remote thermostat for a baseboard unit, or an on-off switch for a kick-space heater.

⚠ **CAUTION** *Do not install baseboard heaters directly under wall receptacles, where cords could accidentally come in contact with the heating element.*

 **TOOLS**

Electronic stud finder
Keyhole saw or saber saw
Electric drill with $\frac{3}{4}"$ extension bit
Fish tape
Wire stripper
Cable ripper
Screwdriver
Pliers
Hammer
Cold chisel

 **MATERIALS**

Electrical cable
Cable staples
Bar hanger
Protector plate
Wire caps
2 x 2
Wallboard nails
Joint compound
Wallboard tape

 **SAFETY TIPS**

*Wear goggles when hammering nails or chiseling and when sawing.*

# A RECESSED WALL UNIT

## 1. Preparing the way.

◆ Choose a location for the heater between two studs, located with an electronic stud finder.
◆ Saw an opening between the studs the size and shape of the heater housing, with the bottom of the hole 2 feet above the floor. If you are installing the heater in a room above another finished room, as shown at right, cut a temporary access hole in the room below to help you fish the cable down to the basement.
◆ With a $\frac{3}{4}$-inch extension bit, drill a cable hole through any intervening sole plates and top plates.

**ELECTRONIC STUD FINDER**

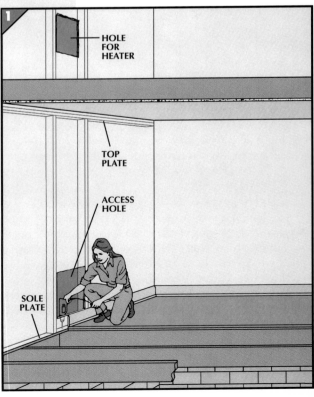

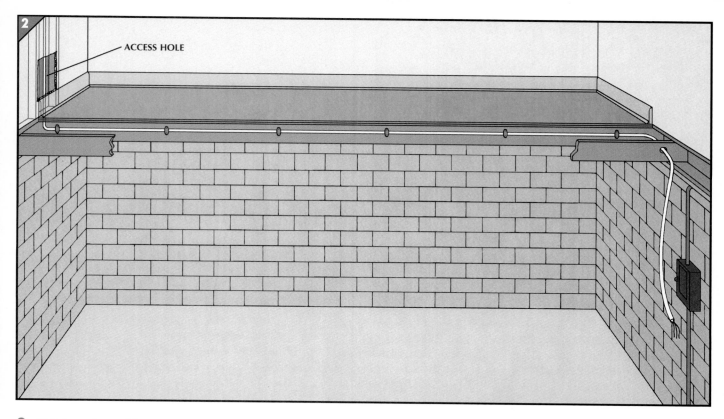

ACCESS HOLE

## 2. Fishing the cable.

◆ Thread fish tape, available at hardware stores, through the hole in the wall and down into the basement. Attach the cable to the tape, and fish up enough cable to reach 2 feet beyond the top of the heater hole.

◆ In the basement, staple the cable to joists at 4-foot intervals along the route to the service panel. If necessary, drill $\frac{3}{4}$-inch holes through the center of the joists near the wall to reach the service panel. Leave about 2 feet of cable at this end for the connection to the panel.

## 3. Clamping the cable to the heater.

◆ Strip about 8 inches of sheathing from the heater end of the cable with a cable ripper. Then use a wire stripper to remove about $\frac{1}{2}$ inch of insulation from the ends of the two conductor wires.

◆ Clamp the two-part connector supplied with the heater around the cable where the wires emerge from the sheathing, threads facing the ends of the wires.

◆ Remove the twist-out from the knockout hole in the top of the heater housing.

◆ Insert the wires and the threaded end of the connector into the knockout hole, then screw the lock nut onto the connector from within the housing *(right)*. Tighten the nut with pliers.

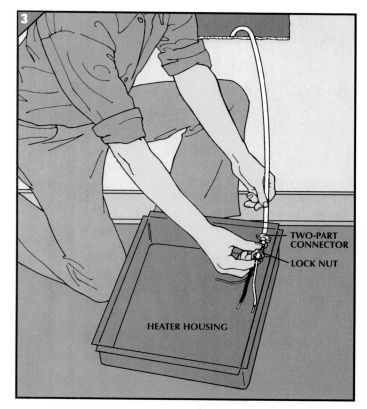

TWO-PART CONNECTOR

LOCK NUT

HEATER HOUSING

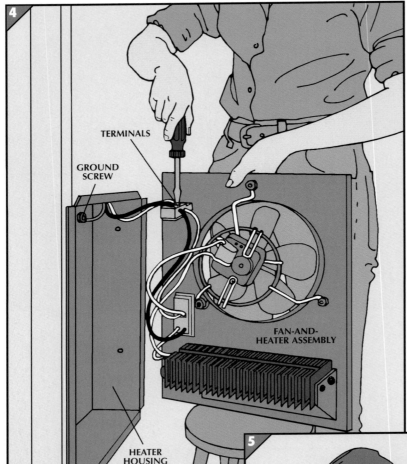

**TERMINALS**

**GROUND SCREW**

**FAN-AND-HEATER ASSEMBLY**

**HEATER HOUSING**

## 4. Making electrical connections.

◆ Screw the heater housing to the studs at the sides of the opening with the screws provided.
◆ Prop the fan-and-heater assembly so that the two screw terminals on top are within reach of the cable wires.
◆ Secure the bare ground wire around the green ground screw at the back of the housing.
◆ For a 120-volt heater, connect the cable wires to the heater wires—black to black and white to white—at the heater terminal. If the heater is a 240-volt model, it makes no difference which wire you connect to each of the terminal screws. However, wrap black tape around the white wire to recode it as a voltage-carrying conductor.
◆ Screw the fan-and-heater assembly into the heater housing.

## 5. Mounting the grille and frame.

◆ If the grille and frame have been packed together, separate them. Screw the grille alone to the top and bottom of the fan-and-heater assembly, making sure that the thermostat shaft protrudes through the center of the hole in the grille. Push the thermostat knob onto its shaft.
◆ Place the heater frame facedown on the floor and press the vinyl gaskets into the inner tracks at the edges of the frame. Hold the frame up to the heater with the gaskets facing the grille, and push the frame into place.

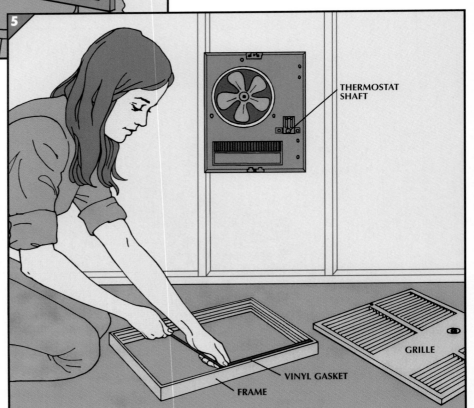

**THERMOSTAT SHAFT**

**GRILLE**

**VINYL GASKET**

**FRAME**

# A CEILING HEATER ON A TIMER

## 1. Cutting a hole for the ceiling box.

Below an unfinished attic, hold an octagonal electric-outlet box against the ceiling where you want to install the heater, and outline the box on the ceiling. Drill a starter hole at each corner, then brace the surface of the ceiling with a block of wood and cut around the octagon *(left)* with a keyhole saw or a saber saw.

If the space above is finished, cut a square hole that spans the distance between two joists and patch it later *(page 71, Step 6)*. Alternatively, cut an octagonal hole as above and insert a screw-type bar hanger, which extends to fit tightly between ceiling joists.

## 2. Installing the box.

◆ Remove one knockout from a ceiling box and attach it to a bar hanger.

◆ If you're using the conventional model, shown at right being installed from an unfinished attic, screw the hanger to the joists so that the edge of the ceiling box is flush with the ceiling. Adjust the screw-type hanger to achieve the same result.

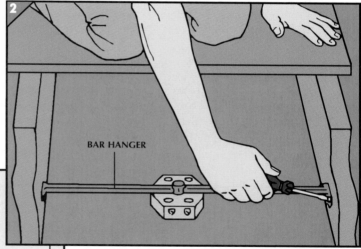

BAR HANGER

WALL-OUTLET OPENING

## 3. Cutting an opening for the switch.

If the attic overhead is unfinished, put the timer switch on whichever wall is convenient. For a finished space above, place the switch between studs that correspond to the ceiling joists that flank the heater. Doing so simplifies fishing cable from switch to heater.

◆ Trace the outline of an outlet box on the wallboard and cut along the outline.

◆ Drill through sole plates and top plates as necessary and fish cable from the service panel to the switch hole *(pages 66-67, Steps 1 and 2)*.

## 4. Fishing cable to the ceiling box.

◆ For an attic that is unfinished (*above, left*), drill a $\frac{3}{4}$-inch hole downward through the top plates between the studs that flank the switch hole. A small hole drilled in the ceiling above the switch hole serves as a helpful guide.

◆ Fish cable from the switch hole to the ceiling box.

◆ If the attic is finished (*above, right*), chisel a slot 4 inches long in the ceiling and 6 inches long in the wall directly above the switch hole, notching the top plates to a depth of $\frac{1}{2}$ inch.

◆ Fish cable from the outlet opening through this access hole to the ceiling box.

◆ Staple the cable to the top plates, and cover it with a metal protector plate, available in hardware stores.

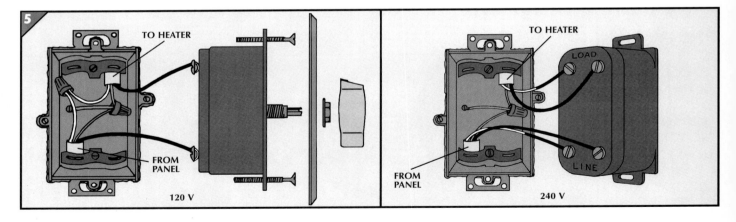

## 5. Installing the switch.

◆ At the switch hole, strip sheathing and insulation from the ends of the cables leading to the service panel and the ceiling fixture. Secure one cable at the bottom of an outlet box, the other at the top, then fasten the box in the hole.

◆ Using a wire cap, connect the two bare cable wires to a short jumper wire of the same diameter, and attach the other end of the jumper to the ground screw in the back of the box. (The jumper is unnecessary with a plastic box.)

◆ For a 120-volt switch (*above, left*), connect the black wires from both cables to the switch terminals. Join the two white wires with a wire cap. For a 240-volt switch, which has four terminals (*above, right*), recode the white wires by wrapping their ends with black tape to indicate that they carry voltage. Then connect the two wires from the service panel cable to the terminals marked LINE and the two wires of the heater cable to the terminals marked LOAD.

◆ Screw the timer switch to the outlet box.

## 6. Patching the hole.

If you cut a large hole in the ceiling, patch it as follows before installing the heater:

◆ Cut two lengths of 2-by-2s a little longer than the exposed portion of the joists. Nail the 2-by-2s to the sides of the joists so that the bottom edges of both pieces are flush.

◆ Cut a wallboard patch to fit the opening, then saw a hole in the patch to fit around the ceiling box. Nail or screw the patch to the 2-by-2s *(left)*.

◆ Fill the seams with joint compound, cover the compound with wallboard tape, and cover that with more joint compound. Let the seams dry, then sand smooth. Apply two more layers of joint compound, feathering the edges, and sand smooth after each layer dries.

## 7. Installing the ceiling heater.

◆ Slip the mounting plate over the heater wires.

◆ Connect cable wires to heater wires with wire caps, black to black and white to white.

◆ Screw the mounting plate and heater to the ceiling box.

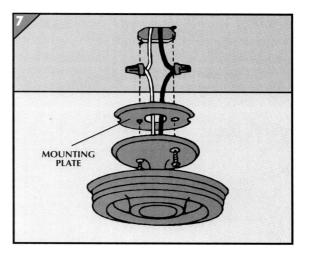

MOUNTING PLATE

# INSTALLING A ROW OF BASEBOARD HEATERS

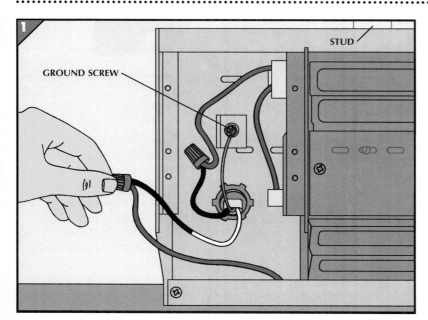

GROUND SCREW

STUD

## 1. Hooking up the first heater.

◆ Remove the baseboard from the wall, then rest the heater on the floor and screw it to studs.

◆ Run a cable to the service panel as shown in Steps 1 and 2 on pages 66 and 67.

◆ Connect the cable ground wire to the heater ground screw.

◆ Inside the heater are three factory-installed wires, usually blue. Two of the wires will be unconnected; attach the cable's black wire to either of these wires. Then recode the cable's white wire black and attach it to the other wire.

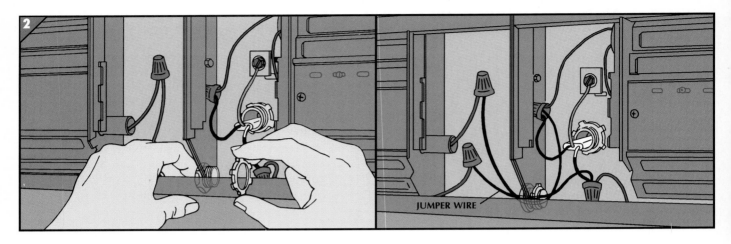

JUMPER WIRE

## 2. Adding a second unit.

◆ Remove the covers and the twist-outs from the knockout holes at the connecting ends of the two heaters.

◆ Screw the new heater to the wall studs, and bolt the two units together through the small holes at the top of the heaters with the nut and bolt provided. Then push a bushing—also supplied—through the matching knockout holes and secure the bushing with its lock nut *(above, left)*.

◆ Join the two loose wires at the far end of the second unit with a wire cap.

◆ Then cut two 10-inch-long black jumper wires and thread them through the bushing. Use the jumpers and wire caps to link the two heaters, making the connections as shown above, right.

# TUCKING A UNIT INTO A KICK SPACE

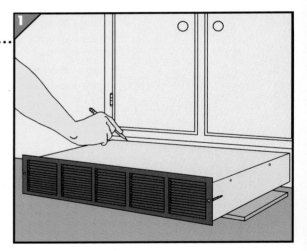

## 1. Positioning the heater.

◆ Rest the heater on a board to center it on the kick-space panel of a cabinet, and outline the heater on the panel. Then saw out the outlined section.

◆ On a nearby wall find a location to which cable can be run conveniently, and cut a hole for an outlet box *(page 69, Step 3)*.

◆ Run cable from the basement to the box opening *(pages 66-67, Steps 1 and 2)* and from the opening to the kick-space hole.

◆ Secure the cables in the wall opening to an outlet box, and install the box in the opening *(page 70, Step 5)*.

◆ Remove the top of the heater, and anchor the cable to the knockout hole in the back *(page 67, Step 3)*.

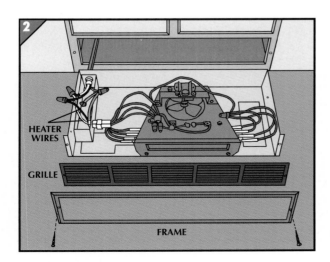

HEATER WIRES

GRILLE

FRAME

## 2. Wiring the heater and the switch.

◆ Connect one insulated cable wire to each of the two unconnected wires in the heater with wire caps, matching insulation colors. Join the bare cable wire, the heater's ground wire, and a short jumper wire with a wire cap, then attach the jumper to the green ground screw in the bottom of the heater.

◆ Screw on the heater top and push the unit into the kick-space opening. Screw the heater frame and the front grille to the cabinet with the screws provided.

◆ Install a standard on-off switch or a timer switch in the outlet box as shown in Step 5 on page 70.

# A Gas Heater on a Wall

When the ductwork of a gas heating system cannot be extended to a room, consider installing a gas-fired space heater, or wall furnace. Mounted on an outside wall, it gives warmth more inexpensively than an electric unit.

Choose a counterflow model that draws air from below the ceiling, heats it, and blows it out near the floor to warm the room evenly. And check ahead of time that the furnace will not require a costly enlargement of the gas-supply line.

**Circulating Air Correctly:** In the wall furnace shown here, two concentric pipes go through a wall to the outdoors. The larger, outer pipe, which attaches to the unit's air-intake collar, brings combustion air to the heater; the inner, smaller pipe connects to a flue collar and carries out exhaust fumes.

**Placing the Wall Furnace:** Position the unit as near to the middle of a wall as possible—provided it is in a spot where gas and electricity can be supplied easily and where the unit can be vented correctly, as stipulated by local code and manufacturer's instructions. Typically, vent pipes must exit the house at least 1 foot above normal snow depth, 3 feet from the nearest corner of the house, and 1 foot from any door or window. Inside the house, maintain the clearances specified by the manufacturer between the heater and nearby furniture, doors, and drapes.

**Power Connections:** You can extend a circuit yourself to provide electricity for the unit *(pages 66-71);* many simply plug into a receptacle. But hire a professional to make the service-panel and gas connections.

If the furnace has a remote thermostat, do not install it next to the unit. Instead, run the thermostat wires connected to the unit to a location on an inside wall and mount the thermostat there *(page 15).* Doing so isolates it from cold that seeps through exterior walls.

 **TOOLS**

Electronic stud finder
Drill with $\frac{1}{8}$" bit
Masonry bit ($\frac{3}{16}$")
Spade bit (1")
Keyhole saw or saber saw
Maul
Cold chisel
Screwdriver
Adjustable wrench
Hacksaw
Carpenter's level

 **MATERIALS**

Hollow-wall anchors
Silicone caulk
Sheet-metal screws (No. 10, $1\frac{1}{4}$")
Screw anchors

 **SAFETY TIPS**

*Work gloves protect your hands from any sharp edges when you handle the wall furnace.*

## 1. Hanging the heater.

◆ Locate two adjoining studs with an electronic stud finder.
◆ Midway between the studs saw a vent hole of the size and height specified for your unit. Remove any insulation from inside the hole.
◆ At the hole's center, drill a $\frac{1}{8}$-inch marker hole through the outside wall.
◆ In the outside wall, cut a vent hole the same size as the inside hole but centered $\frac{1}{4}$ inch below the marker hole. Use a maul and cold chisel for a masonry wall, a keyhole saw or saber saw with other types of siding.
◆ Inside the house, use the vent hole and the back of the heater as guides to position the wall brackets. Fasten the brackets to the wall with hollow-wall anchors.
◆ Find the floor location for the gas-supply pipe and drill a 1-inch hole for it.
◆ With helpers, lift the unit into place, guiding the air-intake collar into the hole and slipping the mounting slots over the wall brackets *(right).*

WALL BRACKETS

VENT HOLE

AIR-INTAKE COLLAR

GAS-PIPE HOLE

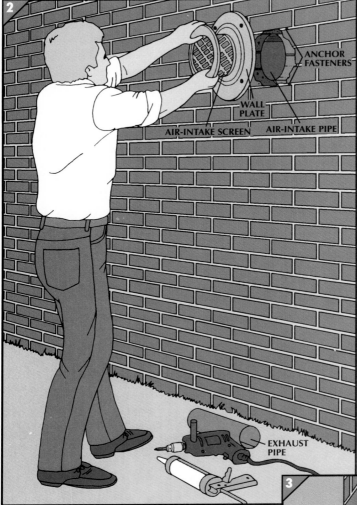

## 2. Installing the intake pipe.

◆ Working outside, fit the air-intake pipe through the vent hole and over the furnace air intake. If the pipe has telescoping sections, adjust them so it protrudes from the wall the distance specified by the manufacturer—usually $\frac{1}{2}$ inch. Otherwise, cut it to length with a hacksaw.

◆ Adjust the pipe so that it inclines slightly downward to the outside—about $\frac{1}{4}$ inch per foot. Check the slope with a carpenter's level; if necessary, enlarge the hole in the outside wall to maintain the slope.

◆ Install the air-intake screen and outer wall plate, which come as a single assembly. For a masonry wall, position the assembly, then mark for the screw anchor positions through the plate onto mortar joints. Remove the plate and drill $\frac{3}{16}$-inch holes at each mark; place the anchors. Spread caulk over the back of the wall plate and attach it to the wall *(left)*. In a nonmasonry wall, caulk the wall plate and fasten the plate to the wall with wood screws supplied with the furnace.

◆ Caulk the plate-to-wall seam.

## 3. Installing the vent cap.

◆ Fit the exhaust pipe over the furnace's flue collar.

◆ Mark the exhaust pipe at the point specified by the manufacturer, typically $3\frac{1}{2}$ inches from the wall. Then trim the pipe at the mark.

◆ Reposition this pipe on the flue collar and fit the vent cap over it *(right).* Then fasten the cap to the wall plate with the nuts and bolts provided with the unit.

# Adding a Hot-Water Convector

Most hot-water heating systems have the reserve capacity to handle one or two additional baseboard convectors. Aside from mounting the unit on the wall, adding one is a plumbing job that consists of diverting hot water from the boiler through the new convector and returning it to the system.

**Tracking the Main:** Before installing a new convector, find the section of the main line, or lines if you have a two-pipe system (below), closest to the planned location of the new fixture—preferably on an exterior wall and under a window. The main may follow the perimeter of the outer walls under the floor joists, or it may run along the center beam of the basement or crawlspace with the risers between the joists.

Once you locate the main, trace the line back to the boiler to determine the direction in which the water flows so you can divert hot water to the baseboard unit. The supply end of the main leads out of the boiler at a higher level than the return end.

**Running New Pipe:** Piping systems vary, but typically the mains are 1 inch in diameter and the risers are $\frac{3}{4}$ inch across. Whether the existing system has pipe of copper or steel, copper is the better material for new installation. It resists corrosion and is easier to work with than steel.

When joining copper pipe to steel, use dielectric unions, even if your local code does not require it; they prevent a corrosive electrolytic reaction between the metals that can cause joints to leak.

 *Older pipe systems may be insulated with material containing asbestos. If you suspect asbestos is present, follow the instructions in the caution on page 48.*

---

 **TOOLS**

Pry bar
Electronic stud finder
Electric drill with $\frac{5}{32}$" bit
Hole saw ($1\frac{1}{4}$")
Hacksaw or tube cutter
Propane torch
Flameproof pad
Carpenter's level
Steel tape measure
Pipe wrenches (2)

 **MATERIALS**

Angle valve
Baseboard T
Pan-head sheet-metal screws ($1\frac{1}{2}$" No. 10)
Solder
Flux
Regular Ts
Venturi Ts
Bleeder valve
Slip couplings
Copper piping ($\frac{3}{4}$" and 1")
Dielectric unions
Pipe joint tape

**SAFETY TIPS**

*When cutting or wrenching pipes, wear gloves. Add goggles if you are working overhead or soldering.*

---

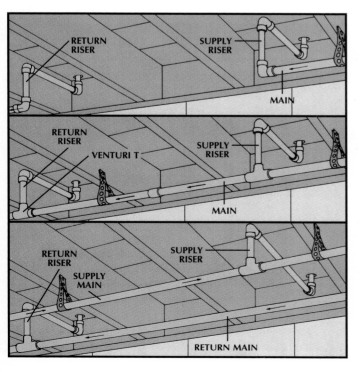

## Hot-water distribution systems.

Three common water-flow patterns for a hydronic system are shown at left. The simplest is called the series loop (top). It circulates all the hot water through each unit in a circuit. This arrangement precludes individual control of convectors—closing a valve at the beginning of a branch would deprive downstream convectors of hot water—and supplies the first convector in a branch with hotter water than reaches the last convector. The result may be uneven heating.

The one-pipe system (center), solves both problems by diverting only a portion of the hot water from the main. One-pipe installations require a venturi T where the return pipe from a convector rejoins the supply line. This special fitting helps pull water through the convector.

In a two-pipe system (bottom), separate supply and return mains ensure that hot water reaches all the units at nearly the same temperature.

# CONNECTIONS FOR A SYSTEM PIPED IN COPPER

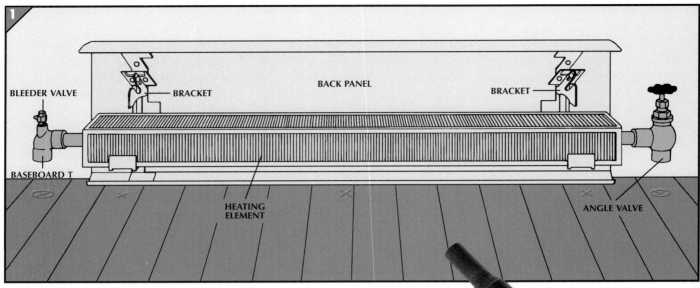

BLEEDER VALVE

BRACKET

BACK PANEL

BRACKET

BASEBOARD T

HEATING ELEMENT

ANGLE VALVE

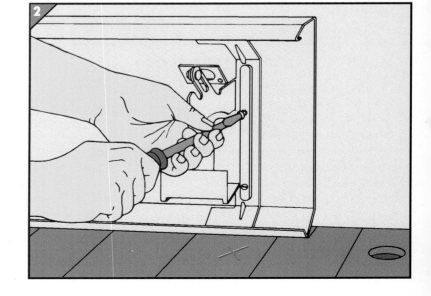

HOLE SAW

## 1. Positioning the convector.

◆ Drain the heating system as described on page 48.

◆ Remove any shoe molding and baseboard from the wall with a pry bar, then locate studs with an electronic stud finder and mark their locations on the floor.

◆ Position the back panel of the convector against the wall. Slide on heating-element brackets and align them with the studs nearest the ends of the unit and at every second stud in between. Seat the element on the brackets.

◆ Hold the back panel against the wall. For a one- or two-pipe system fit on—but do not solder—an angle valve to the unit's supply end and a vented baseboard T to the return end. With a series-loop system, use 90-degree elbows in place of the angle valve and baseboard T.

◆ Mark the floor under the center of each fitting, remove the fittings, and drill marker holes through the floor. Poke a wire through the holes to make sure that they are at least $\frac{1}{2}$ inch from the nearest joist.

◆ Use an electric drill with $1\frac{1}{4}$-inch hole saw *(photo)* to make openings in the floor for the riser pipes.

## 2. Mounting the back panel.

◆ Remove the heating element from the brackets.

◆ Drill $\frac{5}{32}$-inch holes through the top and bottom of each bracket (at the prepunched holes if the bracket has them) and drive $1\frac{1}{2}$-inch No. 10 pan-head sheet-metal screws into the studs.

◆ Set the element on the brackets, without the supply and return fittings.

## 3. Attaching the riser pipes.

◆ Cut two $\frac{3}{4}$-inch copper pipes to match the vertical distance between the convector's supply end and the main under the floor.

◆ Unscrew the larger nut at the angle valve to disassemble it, and unscrew the bleeder valve from the baseboard T. Then solder the pipes to the fittings.

◆ Return the fittings to the heating element, inserting the pipes into the riser holes.

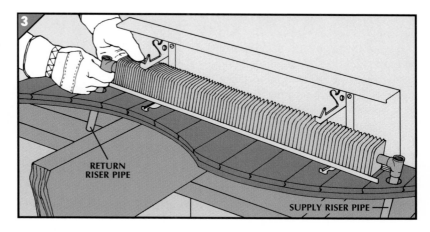

RETURN RISER PIPE

SUPPLY RISER PIPE

RETURN RISER PIPE

MAIN

SUPPLY RISER PIPE

## 4. Cutting the main.

Mark the main opposite each riser pipe. To make space for the Ts that join the risers to the main, cut the main $\frac{1}{2}$ inch outside the marks with a hacksaw or tube cutter, as shown at left.

Follow the same procedure for a series loop, but cut the main $\frac{7}{8}$ inch outside the riser marks to account for one end of each elbow.

In a two-pipe system, mark both the return and the supply mains opposite the corresponding riser pipe, then cut the mains $\frac{1}{2}$ inch outside the marks to account for one end of a T, and 6 inches inside the marks to allow for a length of pipe called a spacer.

## 5. Tying into the main.

◆ Fit an ordinary T onto the main opposite the supply riser pipe, and the return end of a venturi T onto the main across from the return riser pipe. The return end is nearest the narrow end of the venturi *(inset)*.

◆ Trim the removed section of main to fit between the two Ts' internal shoulders.

◆ From this pipe, cut a piece 8 inches long, slide a slip coupling on it, and fit it into one of the Ts. Gently pull one end of the main to the side and fit the longer piece of pipe into the other T.

◆ Align the ends of the two pieces of pipe, then slide the coupling over the joint.

◆ Fit a stub of $\frac{3}{4}$-inch pipe 2 or 3 inches long into each T and use a carpenter's level to be sure the stubs are plumb.

◆ Solder all the joints.

For a two-pipe system, use ordinary Ts for both risers. Trim the spacer cut

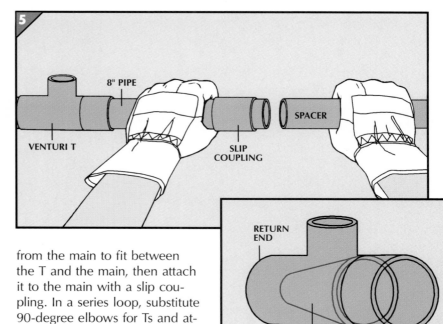

8" PIPE

VENTURI T

SLIP COUPLING

SPACER

RETURN END

VENTURI

from the main to fit between the T and the main, then attach it to the main with a slip coupling. In a series loop, substitute 90-degree elbows for Ts and attach them to the main. No spacer is needed.

### 6. Installing the risers.

◆ For each riser, measure the distance between the riser pipe's center and the center of the pipe stub in the corresponding T. Subtract 2 inches to accommodate two elbows, and cut a piece of $\frac{3}{4}$-inch pipe to fit.

◆ Assemble the pieces as shown at right, then mark each riser pipe $\frac{3}{4}$ inch below the top of the elbow.

◆ Remove the riser pipes from the convector, cut them at the marks, and replace them, slipping the ends into the elbows.

◆ Solder all the remaining joints, including the fittings at the convector upstairs.

◆ Reassemble the angle valve and screw the bleeder valve to the baseboard T.

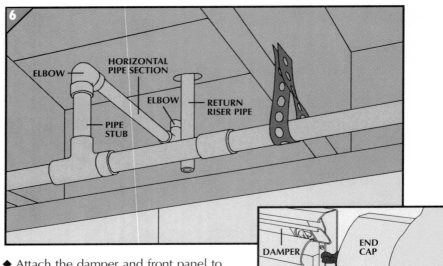

◆ Attach the damper and front panel to the element bracket and cover the ends with caps or wall-trim fittings *(inset)*.

◆ Refill the system *(page 49)*.

# A NEW SECTION IN A STEEL SYSTEM

### 1. Cutting the main.

◆ Prepare the baseboard convector and fit riser pipes to it *(pages 76-77, Steps 1-3)*.

◆ Mark the main opposite the riser pipes.

◆ Saw through the main anywhere between the riser marks.

◆ Unscrew the nearest fitting on each side of the cut, using one pipe wrench to hold the pipe stationary and another to turn the fitting *(right)*.

Do the same for a series-loop system. For a two-pipe system, cut both mains near the riser marks.

### 2. Measuring for the replacement pipe.

◆ Screw a dielectric union onto each end of the steel pipe. Measure the distance between the inside edges of the ring nuts on the dielectric unions *(left)*.

◆ Cut a section of 1-inch copper pipe to that length.

For a two-pipe system, cut a pipe for each main. For a series loop, measure from the inside face of each ring nut to a point opposite the corresponding riser pipe and deduct $\frac{7}{8}$ inch for one end of an elbow. Cut a supply pipe and a return pipe to those lengths.

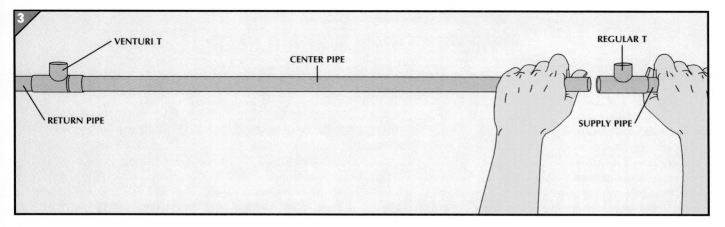

**3. Assembling the replacement section.**

◆ Mark the copper replacement pipe at points opposite the riser pipes.

◆ Measure the distance between the shoulders of a venturi T, and mark the distance on the copper pipe so it is centered over the return riser mark. Repeat the procedure on the supply end for a regular T.

◆ Cut the replacement pipe at all four marks. Discard the two short pieces.

◆ Fit the remaining three pieces to the Ts as shown above.

For a two-pipe system, assemble two new pipe sections using a regular T in both. For a series-loop system, simply attach an elbow to one end of the supply and return pieces.

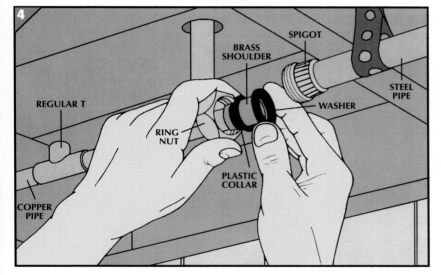

**4. Testing the fit.**

◆ Position the replacement section between the dielectric unions, assembling the unions at each end of the section as shown at left. Tighten the ring nuts on the spigots, using a pipe wrench to keep the steel pipe from turning.

◆ Mark the copper pipe at the brass shoulder of the dielectric union on each end.

◆ Unscrew the ring nuts and take down the copper section. Slide the ring nuts and plastic collars against the T, then solder the shoulders at the marks.

For a two-pipe system, assemble dielectric unions on both replacement sections' ends. With a series-loop system, assemble the dielectric unions where copper pipe meets steel.

**5. The final fit.**

◆ Wrap the spigot threads with pipe joint tape, and when the work is cool, fit the replacement section in place.

◆ Tighten the ring nuts on the spigots.

◆ Fit a $\frac{3}{4}$-inch pipe stub 2 or 3 inches long into the top of each T *(right)* and use a carpenter's level to be sure the stubs are plumb. Solder all the joints.

◆ Install the risers *(page 78, Step 6)* then refill the system *(page 48)*.

Before installing risers in a two-pipe system, solder the joints of both replacement sections. For a series-loop system, solder the elbow joints.

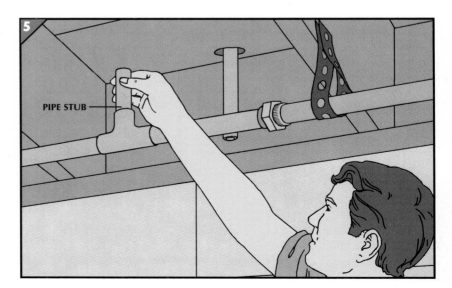

# Stealing Heat to Warm a Room

The fan-coil heater below is an ingenious device that steals hot water from a boiler or, if local codes permit, the house plumbing system. Then, with a water pump and fan powered by a single motor, it uses the hot-water heat to warm a room's air. Fan-coil heaters are available in models that can be flush-mounted against a wall, installed under a cabinet, or recessed between studs (inset, below). You can install one in any room where water pressure is sufficient to keep the heater filled.

Hot-water heat is not free—the BTUs removed by the heater must be restored at the boiler or tank. But the fan and pump, which use little more electricity than a 100-watt bulb, extract heat quite efficiently. Wall-mounted models can provide up to 8,000 BTUH from domestic hot water that is 140° F and as much as 22,000 BTUH from boiler water that is 200° F.

**Plumbing Connections:** Apart from securing the heater in place and making two simple electrical connections (opposite, bottom), the main installation work consists of running $\frac{1}{2}$- or $\frac{3}{4}$-inch copper tubing from the hot-water source to connections on the heater. Gate valves are necessary on these lines so that the heater can be shut off in summer or for servicing.

In an old-style gravity hot-water system, you can often attach the lines to plugged openings in the boiler (called tappings) by removing the threaded plugs and using reducing and transition fittings. On a forced hot-water system, connect to the existing piping (opposite, top), to permit the heater to operate independently of the circulator. T fittings can also be used to hook up to a water heater's existing pipes (opposite, middle).

⚠ CAUTION *Older pipe systems may be insulated with material containing asbestos. If you suspect asbestos is present, follow the instructions on page 48.*

 **TOOLS**

| | |
|---|---|
| Pipe cutter | Wire cutter |
| Hacksaw | Wire stripper |
| Propane torch | Screwdriver |
| Flameproof pad | |

 **MATERIALS**

| | |
|---|---|
| Copper tubing ($\frac{1}{2}$" or $\frac{3}{4}$") | Flux |
| | Power cord |
| Check valve | with ground |
| Gate valves | 2-wire grounded |
| T fittings | cable (No. 18) |
| Nipple | Wire caps |
| Solder | Electrical tape |

🪖 **SAFETY TIPS**

*Wear work gloves and safety goggles when you are soldering copper pipe joints.*

### How a fan-coil heater works.

When installed between wall studs and plugged into an electrical outlet (inset), only the heater's air grilles and control panel are exposed. The working parts are revealed in the cutaway drawing at right. The motor, controlled by an automatic thermostat, simultaneously rotates both the fan and a drive magnet attached to the end of the motor shaft. The drive magnet is centered over a second magnet attached to the pump impeller. The rotating drive magnet causes the impeller to turn, pushing hot water—from either a boiler or a hot-water supply pipe—through finned heating coils. The fan circulates room air over the coils, thus warming the room. Before startup and at the beginning of each heating season, air must be removed from the system's conduits by the opening of a bleed valve, located on the inlet side of the heating coil.

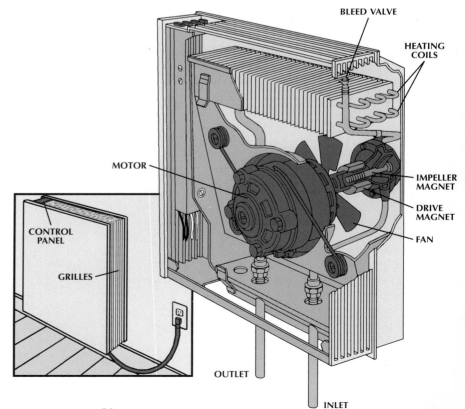

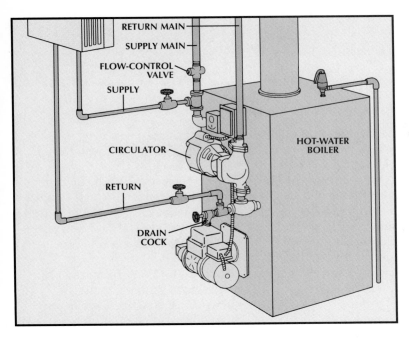

## Tapping a hot-water boiler.

Connect the heater to a forced hot-water system as follows:

◆ Drain the heating system *(page 48)*.

◆ Adapt the methods shown on pages 77 to 79 to insert a T fitting in the supply main on the boiler side of the flow-control valve.

◆ Remove the drain cock from the drain opening on the boiler side of the circulator. Add a T and nipple and replace the drain cock.

◆ Connect a length of copper tubing to the supply T. Attach a shutoff valve to the tubing *(page 51)*, then continue the supply line to the heater.

◆ Similarly extend a return line with a shutoff valve from the drain T to the heater.

◆ Connect the tubing to the heater as specified by the manufacturer.

◆ Refill the system.

## Tapping a water heater.

◆ Turn off power to the heater. Let the water inside cool for a few hours, then drain the heater and the house plumbing system.

◆ Add a T to the water heater's outlet just above the tank, as shown at right, or to the most convenient hot-water pipe *(pages 77-79)*.

◆ Insert another T in the cold-water inlet of the water heater; the dip tube inside the tank will channel the cooler water to the lower part of the tank for reheating.

◆ Extend copper tubing partway from the hot-water T to the fan-coil heater. Add a shutoff valve *(page 51)* and continue the line to the heater.

◆ Attach a line to the cold-water T in the same way, but add a check valve on the fan-coil heater side of the shutoff to prevent any backflow of cold water *(right)*.

◆ Attach the tubing to the fan-coil heater according to the manufacturer's instructions.

◆ Refill the water heater and plumbing system and restore power.

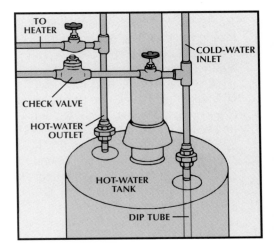

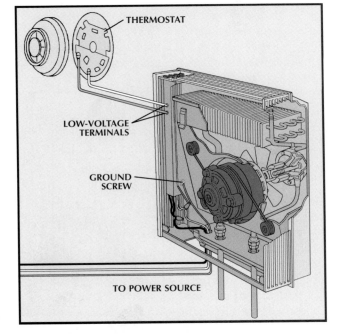

## Wiring the fan-coil heater.

◆ Measure and cut a power cord to run from the heater to the nearest wall receptacle; choose a power cord with a ground wire. (Do not plug in the cord until the heater is ready to run.)

◆ Strip the end of the cord and the ends of each of the three wires inside.

◆ Fasten the ground wire of the power cord to the ground screw on the heater frame.

◆ At the high-voltage junction box inside the heater, connect the power cord's hot wire—normally black—to the two black heater wires. Attach the neutral wire—normally white—to the two white heater wires.

◆ Mount the 24-volt thermostat that is supplied with the unit *(pages 14-15)*. Then run No. 18 two-wire cable from the thermostat terminals to the two low-voltage terminals at the heater. Either wire can go to either terminal.

# Free Heat from the Sun: The Trombe Wall

The Trombe wall—named after French architect Félix Trombe—converts a solid masonry wall into a solar-energy collector that can provide about half the heat for an adjacent living space. Panels of glass, fiberglass, or plastic, mounted in a frame 3 inches out from the house wall, trap solar radiation in a narrow hot-air sandwich, producing a convective air flow through vents during the day and building a reserve of stored warmth that will radiate inward at night.

**Basic Requirements:** Not every wall is suitable. The wall must be solid brick, concrete, stone, or adobe at least 8 inches thick, with no insulation behind it. It must face within 20 degrees of true south, be exposed to direct sun in winter, and be shaded in summer—either by deciduous trees or by an awning. Finally, the masonry must be in good condition, able to withstand temperatures up to 180°F; replace any broken bricks or stones and repair crumbling mortar joints.

**Sizing the Trombe Wall:** For optimum efficiency, the glazed area of the wall should equal about one-third the room's floor space. Adjust the dimensions of the wall to avoid cutting the numerous panels.

To determine the size of the vents, first calculate 1.5 percent of the total Trombe wall area, then divide this figure into an even number of vents. Each vent should be about three times as wide as it is high, with the width not to exceed 20 inches. As shown in the picture on the opposite page, you can use the upper portion of a double-hung window for a top vent and adjust the size of the other vents accordingly. In this example, the right-hand window is left uncovered, providing a view and a fire exit.

**Purchasing the Materials:** Trombe wall panels come in a standard 4- by 8-foot size and are available from solar-equipment dealers. The lighter plastic and fiberglass panels are more popular than glass. Dealers also stock all the necessary framing materials in plastic, aluminum, or wood, along with the appropriate gaskets.

You can save money by buying pressure-treated lumber and cutting the framing pieces. Use neoprene, a black rubber available at hardware stores, for the wall gaskets.

**Building the Wall:** Installation is a fairly simple job but one that requires painstaking measurements. Since glass and plastic panels are quite difficult to trim accurately, an error of as little as $\frac{1}{4}$ inch in the placement of a framing piece can result in an improper air seal. Also, the framing must allow for slight expansion and contraction of the panels caused by temperature change; the allowance is specified by the panel manufacturer.

The measurements given on these pages are for acrylic panels set into a wood frame. If you install a different framing system or another type of panel, refer to the manufacturer's instructions for the correct measurements.

 **TOOLS**

Maul (3-lb.)
Cold chisel
Brick set
Electric drill with carbide-tipped bit
Plumb bob
Circular saw with masonry and plywood blades
Socket wrench
Staple gun

 **MATERIALS**

Wood ($\frac{3}{4}$")
2 x 4s
1 x 4s
$1\frac{1}{2}$" x $\frac{5}{8}$" molding
Mortar
Construction adhesive
Fiberglass insulation
Silicone caulk
PVC pipe ($\frac{3}{4}$")
Expansion bolts
Neoprene gasket ($\frac{1}{8}$")
Hardware cloth
Polyethylene (2-mil)
Duct tape
Finishing nails ($1\frac{1}{2}$")
Wood screws
Aluminum flashing

 **SAFETY TIPS**

*Protect your eyes when you are hammering nails, chiseling, or using a circular saw.*

### How a Trombe wall works.

During the day, the sun heats the air pocket between the panels and the masonry wall, creating an updraft strong enough to push open a flexible damper flap covering the upper vent. Cool air from the room is drawn through the lower vent, pushing open a similar flap, to continue the cycle *(near right)*. At night, the air in the wall space cools, changing the direction of air flow and closing the flaps against screens in the vent holes. But the masonry wall continues to radiate heat it has absorbed *(far right)*.

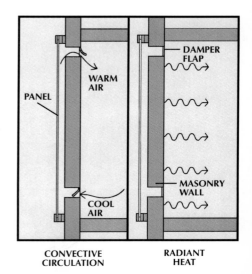

CONVECTIVE CIRCULATION          RADIANT HEAT

**Anatomy of a Trombe wall.**
This Trombe wall is made up of six panels mounted on a frame of doubled 2-by-4s fastened to the masonry wall with expansion bolts. The outer edges of the panels at each end of the wall are sandwiched between $\frac{1}{8}$-inch-thick strips of neoprene gasket, butted against separator strips made of $1\frac{1}{2}$- by $\frac{5}{8}$-inch molding, and covered by

battens made of 1-by-4s. The entire assembly—gaskets, separator strips, and battens—is screwed to the frame.

Interior panel edges butt against separator strips mounted on 2-by-4 mullions. The mullions are offset from the house wall by spacers made of $1\frac{1}{2}$-inch lengths of PVC pipe, through which run expansion bolts that fasten the

mullions to the masonry. The joints between interior panel edges are covered by 1-by-4 battens; screws driven through the centers of the battens into the separator strips anchor the joints firmly. The unvented window is surrounded by a frame constructed of doubled 2-by-4s in the same manner as the perimeter framing, and the plastic panels are cut to fit around it.

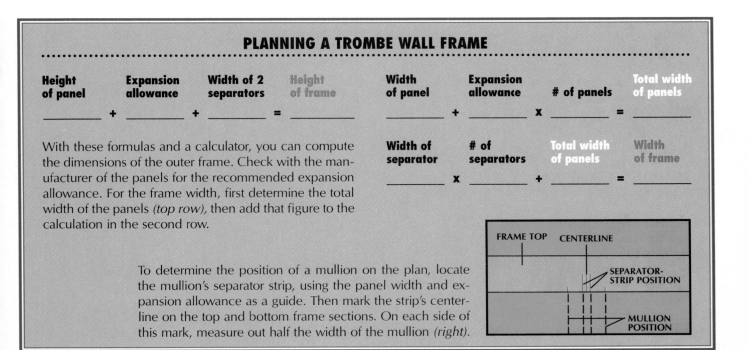

## PLANNING A TROMBE WALL FRAME

| Height of panel | | Expansion allowance | | Width of 2 separators | | Height of frame |
|---|---|---|---|---|---|---|
| _____ | + | _____ | + | _____ | = | _____ |

| Width of panel | | Expansion allowance | | # of panels | | Total width of panels |
|---|---|---|---|---|---|---|
| _____ | + | _____ | x | _____ | = | _____ |

With these formulas and a calculator, you can compute the dimensions of the outer frame. Check with the manufacturer of the panels for the recommended expansion allowance. For the frame width, first determine the total width of the panels *(top row),* then add that figure to the calculation in the second row.

| Width of separator | | # of separators | | Total width of panels | | Width of frame |
|---|---|---|---|---|---|---|
| _____ | x | _____ | + | _____ | = | _____ |

To determine the position of a mullion on the plan, locate the mullion's separator strip, using the panel width and expansion allowance as a guide. Then mark the strip's centerline on the top and bottom frame sections. On each side of this mark, measure out half the width of the mullion *(right).*

FRAME TOP  CENTERLINE

SEPARATOR-STRIP POSITION

MULLION POSITION

# CONSTRUCTING THE VENTS

**1. Breaking through the brick.**
◆ Mark the vent locations on the outside wall. Then cut into the mortar joints along a vent outline with a 3-pound maul and a cold chisel. Try to remove whole bricks within the outline in one piece, but break them if you have to.
◆ To split cleanly through bricks that are crossed by the outline, score them along the outline with a brick set *(photograph)*, then cut along the scored line with the brick set or cold chisel.
◆ Repeat for every vent.
◆ After the outer course has been removed, drill a hole through the center of each vent with a carbide-tipped bit.
◆ Working from inside the house, use the holes as reference points to mark the vent outlines, and remove the inside bricks in the same manner.

**2. Installing the sleeve.**
◆ For each vent, construct a box of $\frac{3}{4}$-inch wood, the same depth as the vent and dimensioned to fit snugly inside it.
◆ Slip the boxes into the vents and fill in the gaps around the hole with a $\frac{1}{2}$-inch-wide line of mortar *(left)*.
◆ While you finish the Trombe wall, stuff each vent with towels or some other material to keep out insects.

# INSTALLING THE FRAME

**1. Establishing string guides.**
◆ At one of the upper corners of the Trombe wall location, drill a 1-inch-deep hole into a mortar joint and tap in a nail at least $4\frac{1}{2}$ inches long; make sure it is perpendicular to the wall and projects approximately $3\frac{1}{2}$ inches.
◆ Mark the height of the frame on the string of a plumb bob and suspend the plumb bob from the nail. Then mark the wall for the lower corner.
◆ Measure the length of the wall along the mortar joint holding the nail, and mark the other upper corner. Hold the plumb bob against that corner, and mark the corner below it.
◆ Check that the four corners form a perfect rectangle by measuring the diagonals between them, adjusting the marks as necessary until the diagonals are equal.
◆ Drill pilot holes and drive nails into the three marked corners. Tie a string near the head of one nail and run the string taut around the others, fastening it to the first nail.
◆ At 1-foot intervals, measure the distance between the string and the wall *(right)*, and slide the string along the nails until the minimum distance is exactly 3 inches.

## 2. Framing the perimeter.

◆ If you plan to paint the Trombe wall, do so with a dark, flat exterior latex paint.

◆ To frame the perimeter, assemble doubled 2-by-4s, leaving 3-inch gaps in the outer 2-by-4s of the top and bottom sections for the mullion joints. Hold the lower edge of the bottom frame section to the guide string, and at one point where the inner 2-by-4 touches the brick, drill through the frame and into the brick.

◆ Apply construction adhesive to the wall side of the inner 2-by-4, and bolt the frame section to the wall with an expansion bolt. Add other bolts at 3-foot intervals; remember that mullion ends will also need bolts. Use shims, if necessary, to hold the face of the outer 2-by-4 flush with the guide string *(right)*.

◆ Attach the top section, then butt the sides against the top and bottom; frame unvented windows similarly. Trim the shims, stuff fiberglass into any gaps between the frame and the wall, and seal the perimeter with silicone caulk.

## 3. Affixing the mullions.

◆ Drill mounting holes in the mullions as for the perimeter framing. Position a mullion in the gaps in the top and bottom frame sections, and bolt it to the wall *(right)*. Install the remaining mullions in the same way.

◆ At 3-foot intervals along each mullion, hold a piece of $\frac{3}{4}$-inch-diameter PVC plumbing pipe against the house wall; mark where the pipe touches the inside face of the mullion *(inset)*.

◆ Cut the pipe at this mark. Then drill a bolt hole through the center of the mullion and a corresponding hole in the wall. Insert an expansion bolt through the mullion and PVC spacer and into the wall. Tighten the bolt.

# ADDING THE PANELS AND FLASHING

## 1. Assembling the separator strips.

◆ Drill $\frac{1}{8}$-inch-diameter weep holes through the separator strips that will be fastened to the bottom frame section; usually one to three weep holes per panel is sufficient, but check the panel manufacturer's instructions for recommended intervals.

◆ Sandwich a gasket between each bottom separator strip and the bottom frame, and, holding the bottom edge of the strip flush with the bottom edge of the frame, nail the strip to the frame with $1\frac{1}{2}$-inch finishing nails every 6 inches. Continue in this fashion, attaching separator strips and gaskets to the sides and top of the frame, but do not drill weep holes into these strips.

◆ In the same manner, attach gaskets and separator strips along the centers of the mullions. As each raised frame of separator strips is completed, check its interior dimensions to be sure a panel will fit into it.

## 2. Securing the panels.

◆ With a helper, hoist an end panel into place within its frame of separator strips. While your helper stabilizes the panel, hold another gasket against the edge of the panel, and top the gasket with a batten. Be sure that the gasket lines up with the outer edge of the batten and that the outer edge of the batten lines up with the outer edge of the frame.

◆ Screw through the batten into the separator strip every 2 feet, using 3-inch wood screws. Continue installing panels in this fashion until all the panels and battens are in place. The finished construction should look like the cross section shown in the inset.

◆ When panels must be cut, as for fitting around a window, use the tool recommended by the manufacturer; for most plastic panels, this will be a circular saw fitted with a plywood-cutting blade. Apply masking tape to panel areas along which the saw will travel, to prevent the saw base from scratching the panel.

## 3. Installing the flashing.

◆ At the mortar joint immediately above the top of the Trombe wall, use a circular saw with a masonry blade to rout a $\frac{3}{4}$-inch-deep groove along the entire length of the wall.

◆ Fashion—or have a fabricator fashion—strips of aluminum flashing wide enough to fit into the groove, project over the Trombe wall, and bend over the wall's front edge about 1 inch.

◆ Fill the groove with mortar and insert the flashing edge; hold the flashing there until the mortar stiffens.

# FINISHING THE VENTS

### Adding the draft flaps.

◆ Working inside the house, remove the rigid foam insulation from the vents. Cut a piece of the wire mesh called hardware cloth to fit the exterior dimensions of the upper vents, and staple the mesh to all four edges of the vent. Repeat for each upper vent.

◆ Cut a 2-mil polyethylene flap $\frac{3}{8}$ inch longer and wider than the interior dimensions of the vent. Bend duct tape over the top $\frac{1}{2}$ inch of the polyethylene for reinforcement, and staple the flap to the top edge of the vent *(below, left)*. Repeat for each upper vent.

◆ For lower vents, cut the flaps $\frac{3}{4}$ inch narrower than those for the upper vents. Staple each flap to the edges of the vent first, then add the mesh *(below, right)*.

◆ Frame the vents with molding of your choice; here, clamshell molding is used *(inset)*. Fasten the outer portion of the molding to the wall surface with construction adhesive, and nail the inner portion to the edges of the vent with $1\frac{1}{2}$-inch finishing nails.

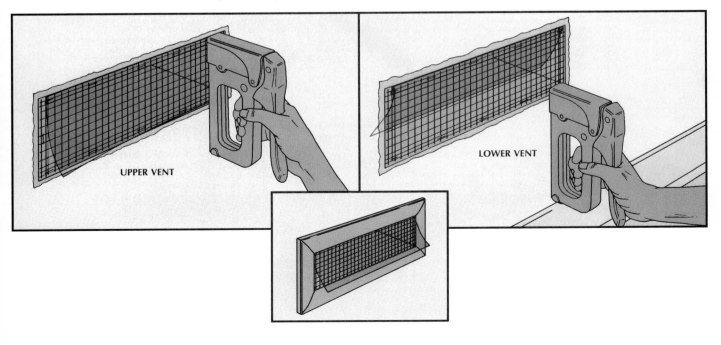

UPPER VENT

LOWER VENT

# Hot Water from Sunshine

Tapping the sun to heat a swimming pool or to supplement a home furnace or water heater can lead to worthwhile fuel savings. How much money you can preserve depends largely on where you live.

For example, the Energy Efficiency and Renewable Energy Clearinghouse in Merrifield, Virginia, a nationwide energy information and technical assistance service, estimates that in the sunny Southwest a solar hot-water system like the one shown here can trim 90 percent per year from the hot-water bill for an average-size single-family home. At that rate, a system would pay for itself in about 10 years, sooner if you install the equipment yourself—a fairly uncomplicated plumbing and electrical job.

**An Appropriate Site:** For a solar system to be economical, the collectors—water-heating panels—must be exposed to the sun from 9 a.m. to 3 p.m. Suitable locations are a backyard, a flat roof, or a pitched roof if one side faces south and the pitch is within 15 degrees of the latitude of the place where you live—a figure you can get from an atlas. The angle of any pitched roof will usually be within acceptable limits.

**The Right Pieces:** Consult a solar-heating dealer for the number and size of collectors to buy and the capacity of the drain-back tank you'll need as a reservoir for the water draining from the collectors when the system is idle. A dealer can also guide you to the right pumps and other components, including a tempering valve to prevent scalding. Plan on supplying power to the system with a new 15-amp electrical circuit from the service panel.

## TOOLS

Crosscut saw
Electric drill with
  bits ($\frac{1}{4}$", $\frac{3}{4}$")
Chalk line
Propane torch
Flameproof pad
Tube cutter
  ($\frac{1}{2}$" to $1\frac{1}{4}$")

## MATERIALS

Common nails (3")
2 x 4s
Aluminum angle (1")
Silicone sealant
Lag screws ($\frac{1}{4}$" x 4")
  and washers
Copper pipe ($\frac{3}{4}$"),
  couplings,
  reducers, and
  caps
Lead-free solder
Sheet-metal screws
  (1", self-tapping)
Rubber pipe
  insulation
Lag shields
  ($\frac{1}{4}$", long)
Electrical cable
Hose bib
Outlet box
Grounded
  receptacle
Junction box
Wire caps

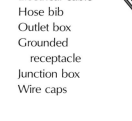

### SAFETY TIPS

*Rubber-soled shoes provide sure footing on a roof, but use ladders and ladder hooks (opposite) if you feel insecure on a pitched roof. Wear goggles to shield your eyes when drilling. Add work gloves when soldering.*

### Anatomy of a solar heater.
The solar water-heating system below preheats water for a conventional water heater. To do so, a pump (1) circulates water through solar collectors on the roof (2). The heated water then passes through a drain-back tank (3) to a heat exchanger (4), where heat in the collector water is transferred to water in the solar water tank (5), which is also circulated by a pump (6). This solar-heated water replaces hot water drawn from the conventional water heater (7), which supplies additional heat as needed. Cold water from from the main replenishes the solar tank.

Strategically placed thermal sensors prevent unintentional cooling of the water in the solar tank by turning off the pumps when the temperature of the water from the solar collectors falls to within 5°F of the water in the solar tank. When the system is off, water drains from the collectors into the drain-back tank for storage until the pumps restart. Valves in the collector circuit allow for filling and draining the system; others permit bypassing it.

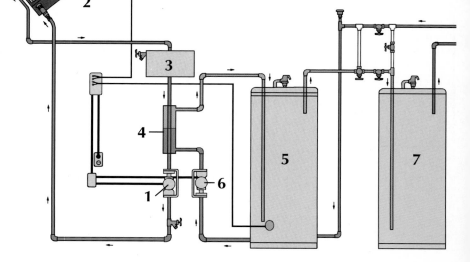

# INSTALLING SOLAR COLLECTORS AND THEIR PLUMBING

### 1. Locating the collectors.
◆ Inside the attic, find a rafter suitably positioned to anchor one end of the line of collectors. Drive a 3-inch nail through the roof next to the rafter about 6 inches below where the bottom of the collector will rest. Note the spacing of the rafters—usually 16 or 24 inches from center to center.
◆ On the roof, find the protruding nail and measure $\frac{3}{4}$ inch to the center of the rafter. With chalk, mark the center of this rafter and others along a horizontal line the length of the row of collectors.
◆ Add 1 foot to the height of the collectors and use the distance to mark the centers of the same rafters higher on the roof *(left)*. Remove the nail and patch the hole.

### 2. Installing collector supports.
◆ From pressure-treated 2-by-4s, cut two 4-inch blocks for each rafter the collectors will span.
◆ Ascend to the roof. Six inches above the lower row of rafter center marks, use a nail to attach the end of a chalk line where the collector inlet will be. Then anchor the line 10 feet away and $2\frac{1}{2}$ inches higher; this slopes the collector $\frac{1}{4}$ inch per foot for draining the system. Snap the line, then remove the nails and patch the holes.
◆ At each rafter, center a block on the chalked line, then lay a support strip of aluminum angle on the blocks and along the chalked line *(inset)*. With a $\frac{1}{4}$-inch bit, drill a pilot hole through the angle, block, and roof into the rafter.

◆ Squirt silicone sealant into the holes, and secure the angle and block to each rafter with 4-inch lag screws and washers. Cover the screwheads with sealant, and lay a bead as well along the top edge of any blocks that have gaps in the shingles beneath them.
◆ As a guide for a chalked line to mark the upper support angle's position, cut a board $\frac{1}{8}$ inch longer than a collector's height, and rest one end on the angle screwed to the roof. At the other end, snap a chalk line to mark the upper angle's position, then fasten it and the remaining blocks to the roof as before.

### 3. Raising the collectors.
◆ Tie a rope securely around a collector, leaving two long rope ends. Using two roof-mounted ladders with hooks as secure footing for you and a helper, haul the collector up to the roof on a third ladder *(above)*.
◆ Set the collector between the aluminum support angles.
◆ Similarly, haul up the second collector and place it next to the first one.

## 4. Connecting the collectors.

◆ Each collector has a copper manifold nipple top and bottom that corresponds to a similar fitting on its neighbor. Slip a copper coupling onto each of the copper manifold nipples of one collector in a pair, then position the two so that the nipples are about $\frac{1}{8}$ inch apart *(left)*.

◆ Slide the couplings so that they bridge the gaps, and solder them to the nipples.

◆ If the nipples are 1 inch in diameter, solder a 1- to $\frac{3}{4}$-inch copper reducer *(inset)* to the outlet nipple at the highest collector corner and the inlet nipple at the lowest. For $\frac{1}{2}$-inch nipples, use $\frac{3}{4}$- to $\frac{1}{2}$-inch reducers flipped around to step up the pipe size to $\frac{3}{4}$ inch.

◆ Solder copper caps over the two other nipples' mouths.

⚠️ **CAUTION** *Here and elsewhere, when soldering, protect the roof with a flameproof pad.*

## 5. Securing the collectors.

◆ Attach the collectors to both supporting angles with 1-inch self-tapping sheet-metal screws that are driven at 1-foot intervals *(right)*.

◆ Push an electronic thermal sensor between the rubber boot and the inlet nipple for the array, and let it hang loosely inside the collector *(inset)*.

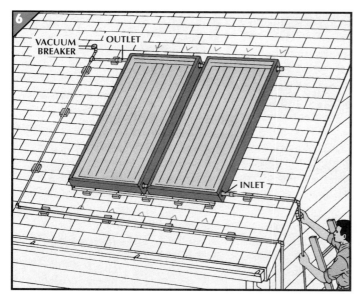

## 6. Attaching the pipes.

◆ Cut more 2-by-4 blocks, and secure one under the outlet. Then extend $\frac{3}{4}$-inch copper pipe from the outlet to the next rafter, and insulate it with rubber pipe insulation.

◆ Connect a T to this pipe, and install a vacuum breaker to a short length of pipe and an elbow connected to the top of the T. The breaker should rise perpendicular to the roof.

◆ Affix blocks and lay and insulate pipe from the T to the bottom of the collectors, then across the roof on an angle sloping along the base of the collectors to the roof edge.

◆ Install blocks and lay and insulate pipe in the same way from the inlet nipple.

◆ Extend both pipes down the side of the house *(left)*.

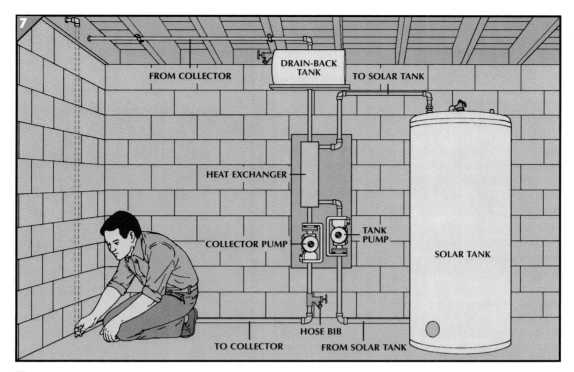

**7. Connecting indoor components.**

◆ Drill two $\frac{3}{4}$-inch holes through the band joist in your basement, extend pipes through the holes, and seal them with silicone sealant.

◆ Mount the drain-back tank on a sturdy shelf near the ceiling.

◆ Extend the outlet pipe to the drain-back tank, strapping it to joists as necessary. Solder a copper adapter to it, then screw the adapter to the inlet atop the drain-back tank.

◆ Secure a piece of $\frac{3}{4}$-inch plywood to the wall with 2-inch lag screws and lag shields.

Then mount the heat exchanger and pumps on the plywood. Orient the pumps so that one sends water to the collector inlet pipe and the other takes water to the top of the solar tank.

◆ Link the solar tank through its pump to the heat exchanger.

◆ Using another adapter, join the drain-back tank to the heat exchanger, then connect the exchanger to the collector pump.

◆ For draining the system, install a hose bib— a common threaded faucet—in the inlet pipe below the pump.

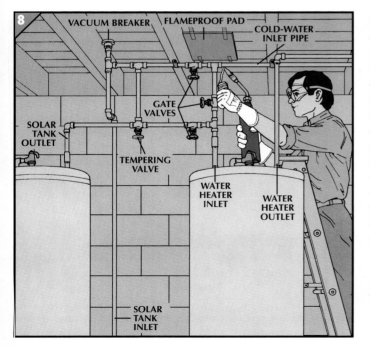

**8. Hooking up the water heater.**

◆ Turn off the cold-water supply to the water heater, and remove the elbow and pipe that connect the two.

◆ Attach an adapter to the solar tank's outlet, and run $\frac{3}{4}$-inch pipe through an elbow and toward the water heater inlet. En route, install a tempering valve, a gate valve, and a T directly above the water heater inlet.

◆ Run pipe from the inlet, through the T and a gate valve, to a T in the cold-water supply line that replaces the elbow removed earlier.

◆ Extend the cold-water supply from this T, through a gate valve to another T above the tempering valve, then onward to a T above the solar tank inlet.

◆ Install a vacuum breaker atop this T, and link the tempering valve and solar tank inlet to the corresponding Ts.

⚠ **CAUTION** *Remove the handles and stems from valves before soldering them in place, to prevent the washers from melting.*

# PLACING THE SYSTEM IN OPERATION

## Wiring and filling the system.

◆ Mount the control panel near the pumps.

◆ Run cable for a new 15-amp circuit from the service panel to a point nearby, and install an outlet box with a grounded receptacle.

◆ Place an electronic thermal sensor at the bottom of the solar tank. To do so, open the access door near the bottom of the tank and slip the sensor behind the clip provided so that it touches the tank *(inset)*. Close the door.

◆ Run wires from the collector sensor and the solar tank sensor to the corresponding terminals on the wall-mounted control panel.

◆ Mount a junction box near the pumps. Inside, wire them to each other with wire caps and to a cable that runs to the control panel. There connect the cable to the control panel wires labeled CONTROLLED OUTPUT.

To charge the system with water, connect a hose from a cold-water faucet to the hose bib below the collector pump; open this valve. Open the valve on the drain-back tank, attach another hose, and extend it to a floor drain.

◆ Run water into the system until a steady stream of water comes out of the drain-back tank valve. Turn off the water and close the valve below the collector pump.

◆ When the flow from the hose attached to the drain-back tank stops, close the drain-back valve. The system has reached its proper level.

◆ Plug the control panel cord into the receptacle to start the pumps.

◆ Check for leaks in the pipes that lead to and from the collectors, including the nipples at the top and bottom of each. Resolder joints as necessary, after unplugging the control panel in order for the system to drain.

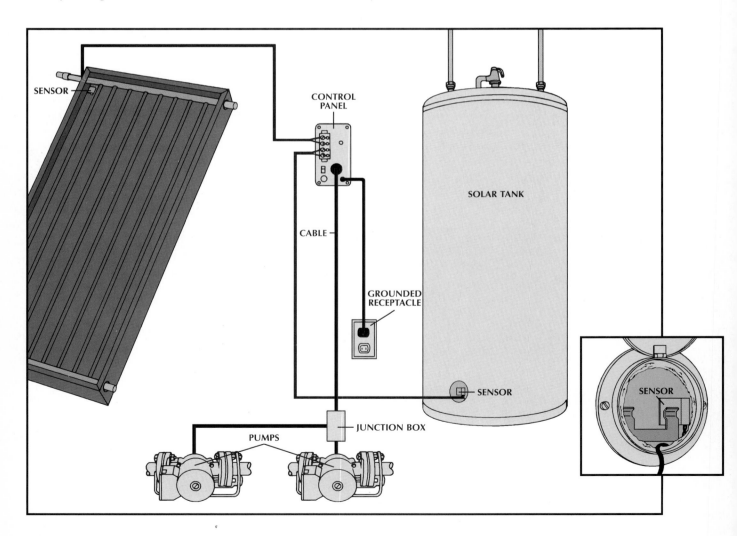

# Installing a Solar Pool Heater

With considerably less work than required to set up a sun-powered water-heating system for the house *(pages 88-92)*, you can install solar collectors as the sole heat source for a swimming pool or as auxiliary heaters to save fuel.

Aside from mounting the collectors on the roof, you need only install a valve and Ts to divert the water through the collectors on its way to the pool. The work is simplified because a solar pool heater uses chlorinated polyvinyl chloride (CPVC) pipe, which is easier to work with than copper. Also, a swimming pool already has a pump to circulate the water.

**About the Panels:** Plastic pipe with no glass casing, collectors for warming a swimming pool are comparatively lightweight but less efficient. The simple, inexpensive design heats water to about 80°F instead of the 125°F of household water.

The number of collectors needed depends on where you live. In Zone 1 of the map on page 125, the collectors should cover an area one-third the size of the pool; in Zone 2, one-half the pool size; and in Zone 3, three-fourths the pool size.

 **TOOLS**

Chalk line
Electric drill with $\frac{3}{16}$" bit
Hacksaw

 **MATERIALS**

Silicone sealant
Solar collectors
One-hole straps
Lag screws ($\frac{1}{4}$" x 2") and washers
CPVC primer and cement
CPVC pipe and fittings
Ball valve

**SAFETY TIPS**

*Although rubber-soled shoes provide sure footing on a roof, use ladders and ladder hooks if you feel insecure on a pitched roof. Wear goggles when drilling. Protect your hands with rubber gloves when cementing CPVC pipes.*

ONE-HOLE STRAP

## 1. Anchoring collectors to a roof.

◆ Locate and mark rafters as in Step 1 on page 89, and snap a slanted chalk line along the top marks.
◆ Along the line drill a $\frac{3}{16}$-inch pilot hole in each rafter, then fill the holes with silicone sealant.
◆ Pull the first collector into place, and fasten one-hole straps to the rafters with 2-inch lag screws and washers. Raise the other collectors to the roof and hook them together with couplings and CPVC cement, then strap them to the roof *(above)*.
◆ Run CPVC pipe from the inlet and outlet manifolds to the swimming pool pump-and-filter system.
◆ Cap manifold openings that will not be connected.

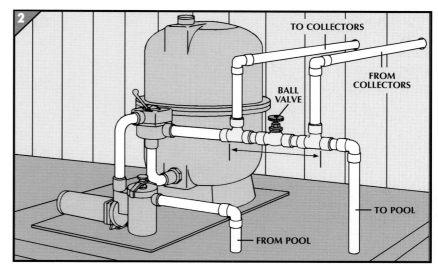

TO COLLECTORS
FROM COLLECTORS
BALL VALVE
TO POOL
FROM POOL

## 2. Hooking up the collectors.

◆ Attach a T to each side of a ball valve with pipe to create a valve-and-T assembly. Measure between the Ts *(arrow)*, and add $\frac{1}{4}$ inch.
◆ From the pipe that is leading to the pool—before the pipe enters an existing heater, if there is one—cut a section to this length. Replace the section with the valve-and-T assembly *(light blue)*.
◆ Using elbows, join the collector inlet to the T nearest the pump and the outlet to the T nearest the pool. By closing the valve, you can increase the flow of water to the collectors—and the pool temperature. Or you can install thermal sensors similar to those for the hot-water system shown on page 92.

# Cooling Your House

Central air conditioning may seem the logical choice to fend off uncomfortably hot weather, but there are other options available as well. For a house without an existing duct system, room air conditioners would prove more economical to install. Or if you live in a region where excessive humidity is not a problem, a whole-house fan or evaporative cooler might suffice.

Cleaning the coil fins on a window air conditioner →

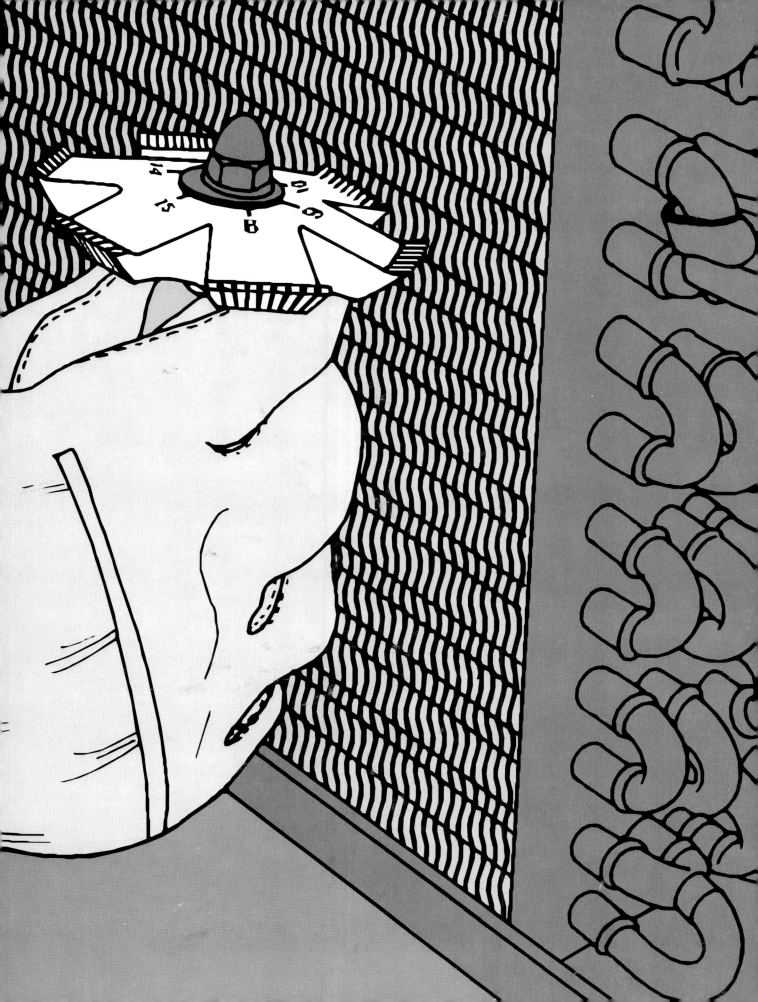

# Lowering Cooling Costs with an Attic Fan

**W**hen summer comes and the temperature rises, an attic fan can be a big money-saver. Air in an unvented attic may reach 150°F, making your home's air conditioner run continuously to hold the temperature in the living quarters at 78°F. A fan can reduce the attic temperature to 95°F and cut your air conditioning bill by as much as 30 percent.

**A Roof Fan:** The most efficient type of attic fan is one that expels hot air through the roof. The model shown on these pages is housed in an assembly mounted over a hole in the roof; its assembly includes flashing to divert runoff water, a cover to protect the fan motor from rain and snow, and a screen to keep out birds and large insects. Installing the unit is simple and safe in an asphalt-shingle roof but may break wood or slate shingles and is impractical in a metal roof.

As part of the installation, you may need to add soffit vents at regular intervals under the eaves on both sides of the house to replace the heated air expelled by the fan. Use the formula below to calculate the vent area required.

**A Gable Fan:** An easy-to-install alternative to a roof unit is a fan mounted inside a vented gable. It creates a flow of air across the attic from a vent in the opposite gable. As with a roof unit, an adjustable thermostat turns the fan on and off at preset temperatures.

 **MATERIALS**

Extension ladder
Ladder hooks
Tape measure
Hammer
Linoleum knife
Saber saw

Electric drill with
 $\frac{3}{8}$" bit
Spade bit ($\frac{3}{4}$")
Pry bar
Screwdriver
Fish tape
Wire cutter
Wire stripper

 **TOOLS**

Roofing cement
Galvanized common nails ($2\frac{1}{2}$")
Lumber (2 x 4)

Fan thermostat
2-wire grounded
 cable (No. 12)
Screw-on wire caps
Soffit vents with
 mounting screws

 **SAFETY TIPS**

*Wear rubber-soled shoes when on the roof, and use a ladder with ladder hooks over the ridge on a roof with a slope of more than 4 inches in 12. Use goggles, dust mask, and ear protectors when sawing, and goggles when nailing. Wear a hard hat in the attic for protection against projecting nails. Wear heavy work gloves when using a linoleum knife.*

# CALCULATIONS FOR FAN AND VENT CAPACITY

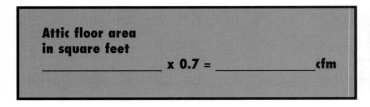

**Attic floor area
in square feet**

_____ x 0.7 = _____ cfm

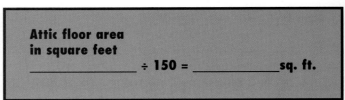

**Attic floor area
in square feet**

_____ ÷ 150 = _____ sq. ft.

## Fan capacity.
To determine the minimum capacity for a roof or gable fan, measured in cubic feet per minute (cfm) at "static air pressure," use the formula above. Add 15 percent for a dark roof, which absorbs more heat than a light one. If a fan is rated at "free air delivery," subtract 25 percent from its cfm rating to get its true capacity.

## Estimating net vent area.
Use the formula above to determine the "net vent area" required for inlet vents. This formula is for vents that offer no more resistance to airflow than $\frac{1}{2}$-inch screening. Vents with metal louvers call for $1\frac{1}{2}$ times the net vent area; those with wood louvers call for twice as much. Most vents are 8 by 16 inches or smaller; install as many as needed to make up this area.

# FITTING A FAN INTO A ROOF

### 1. Positioning the fan.
◆ Assemble the fan and carry it up to one end of the roof, near the ridge on the side where you will install it.
◆ Using a piece of wood as a guide, position the assembly so the cover's top is level with the ridge *(left)*. Measure from the ridge to the center of the fan.
◆ In the attic, on the same side of the roof, locate the two central rafters (midway between gables). On a line halfway between them, mark off the ridge-to-fan distance you measured earlier. Drive a long nail up through the roof at this point.

### 2. Removing shingles and underlay.
◆ On the roof surface, locate the marker nail and use it as a center to mark a circle on the shingles, about 4 inches wider than the fan opening specified by the manufacturer.
◆ Use a linoleum knife to cut along the marked circle to the wood or plywood sheathing underneath.
◆ Remove any nails within the circle that hold the shingles and underlay in place, then pull up and discard the roofing inside the circle.

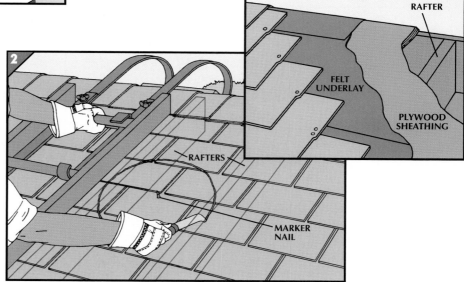

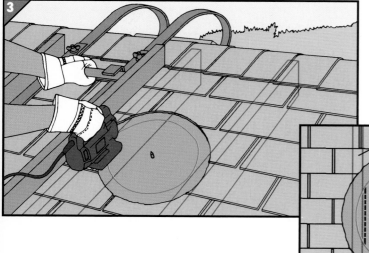

### 3. Cutting through the sheathing.
◆ Using a piece of string tied around the marker nail at one end and a pencil at the other, mark a circle that is the size of the specified fan opening.
◆ Drill a hole inside the circle and near the line, then start a saber saw in this hole and cut entirely around the marked circle.
◆ If the hole specified by the manufacturer is wider than the interval between the rafters, do not cut through them; saw along their inner edges *(inset)*.

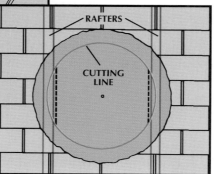

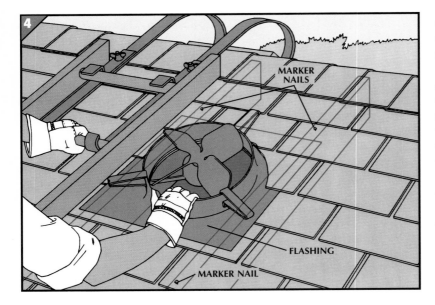

## 4. Installing the fan housing.

◆ Position the fan over the hole and drive four marker nails partway into the roof and rafters below: two just above the top edge of the flashing and two about 6 inches below the bottom edge.

◆ Apply a liberal coat of roofing cement to the exposed sheathing and to the underside of the fan flashing.

◆ Remove the fan cover so that you can see through the fan assembly, then slip the flashing up under the shingles until the housing hole lies directly over the roof hole. With a pry bar remove any shingle nails that obstruct the flashing.

◆ Using the marker nails as guides, drive $2\frac{1}{2}$-inch galvanized common nails through the flashing and into the rafters below at 3-inch intervals. Remove the marker nails.

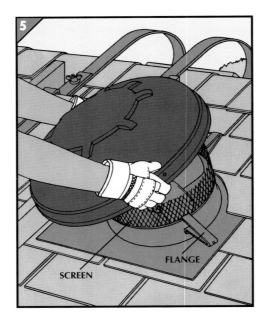

## 5. Installing the screen and cover.

◆ Slide the cylindrical screen into the fan opening and secure it to the hooks on the sides of the opening.

◆ Set the cover over the opening and bolt it to the flanges at the rim of the housing.

## 6. Installing soffit vents.

◆ On the underside of the eave, locate the lookout beams by finding the rows of nails that secure the soffit to the beams.

◆ Hold the vent centered between the outside wall and the edge of the eave, and between two lookout beams. Outline the screened area of the vent on the soffit.

◆ Use a saber saw to cut out the vent opening, insert the vent into the hole from below, and secure it to the soffit with screws through the vent's flange holes.

**7. Wiring the fan.**

◆ In the attic, fasten the fan thermostat to a rafter, with its control dial readily accessible and the temperature-sensing element on its back exposed to the air.

◆ To bring power to the fan from a receptacle on a lower floor, drill a $\frac{3}{4}$-inch hole with a spade bit through the top plates directly above the receptacle. Turn off the power to the receptacle and remove it from the outlet box.

◆ Fish cable from the thermostat to the receptacle through the hole in the top plates; clamp one end of the cable to the thermostat and the other end to the receptacle outlet box.

◆ Using wire caps, connect the black wire of the cable to

the black wire of the thermostat, the white cable wire to the white thermostat wire, and the bare wire to the ground screw.

◆ If the other end of the cable runs to an end-of-the-run receptacle, attach the black and white cable wires to the free terminals. With a wire cap, join the bare ground wire with the existing ground and a short jumper wire. Run the jumper to the receptacle ground screws.

◆ To connect with a middle-of-the-run receptacle *(inset)*, remove a black wire from its terminal and use a wire cap to join it to the new black wire and a black jumper wire *(dashed lines)*; attach the jumper wire to the terminal. Connect the new white wire *(dashed lines)* in the same way. Attach the new ground wire *(dashed lines)* to the other bare wires and to the jumper wire.

# MOUNTING A FAN ON A GABLE VENT

**Over a rectangular vent.**

◆ Position the fan at the center of the vent, and mark the width of the fan cylinder on the headers above and below.

◆ Cut two 2-by-4 supports to fit the vertical distance between the headers, and nail them in place with their inside edges flush with the marks.

◆ Screw the ventilator fan to these supports, and attach the thermostat to a convenient stud nearby. Connect the thermostat to a receptacle *(above, Step 7)*.

◆ If the gable vent at the other end of the attic is too small to serve the fan *(box, page 96)*, add soffit vents *(opposite, Step 6)*.

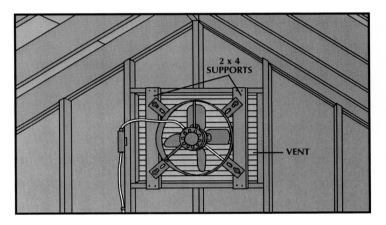

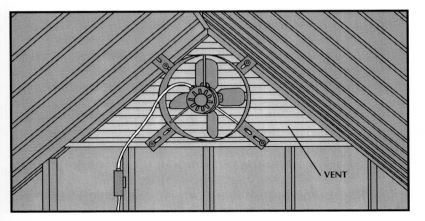

**Over a triangular vent.**

◆ If the fan flanges reach to the vent's wooden frame, screw the assembly directly to the frame.

◆ If the frame is larger than the flange, install vertical 2-by-4 supports *(above)* and attach the assembly to these supports.

# A King-Size Fan to Cool a House

A powerful fan, mounted on an attic's floor joists, can create a pleasant breeze throughout a house. Less expensive to install than central air conditioning, it also costs less to run.

Controlled by a switch downstairs, the fan pulls fresh air in through open windows and drives attic air out through gable and soffit vents. A louvered shutter below closes automatically when the fan is idle, sealing off hot summer air.

**Installation Methods:** Place the fan in a central location, or near a stairwell. Some models can be mounted directly on top of joists. Others require the construction of a supporting wood frame.

If the fan fits between joists, the frame consists of a square that is formed by two joists and two crosspieces nailed between them. More often, the frame must be built as shown opposite.

**Buying a Fan:** To determine how big a fan you need, use the formulas below. Look for a unit with an automatic shutoff for an overheated fan, and a control that shuts down the fan in a fire.

Ask the dealer for the correct attic vent area for your fan and calculate the net area for metal- or wood-louvered vents *(page 96)*. Methods for installing vents are shown on page 98.

 **TOOLS**

Utility knife
Hammer
Electric drill and bits
Keyhole saw
Circular saw or saber saw
Tape measure
Screwdriver

 **MATERIALS**

Cardboard sheet
Joist lumber
Common nails (2", $3\frac{1}{2}$")
1 x 3s

 **SAFETY TIPS**

*Safety goggles protect your eyes when you are hammering, working with power tools, and handling insulation. This material also requires a dust mask to prevent the inhalation of fibers and work gloves to guard against skin irritation.*

## CALCULATING NET AIR VOLUME

| Total floor area in square feet | Average ceiling height in feet | Gross air volume | |
|---|---|---|---|
| _____ | x _____ | = _____ | cubic feet |

| Gross air volume | | Net air volume | |
|---|---|---|---|
| _____ | x 0.9 = | _____ | cubic feet |

The top formula gives the total volume of your house, which the second formula discounts for closets and other spaces that need no cooling. The result is net air volume. In the southern United States, buy a fan capable of moving this quantity of air each minute. Half that capacity is adequate in the North and Canada.

# FRAMING AN OPENING

## 1. Making a template.

◆ Place the shutter, louvers up, on a sheet of cardboard and draw around it with a pencil. Cut along the line with a utility knife in order to make a template.
◆ Drive a nail through the ceiling at the location you have chosen for the fan.

◆ In the attic, remove any insulation that may be covering the nail and set it aside for use in cold-weather months to cover the fan.
◆ Set the template over the nail with one edge against a joist. Mark the corners of the edge on the surface below and drill two small holes at these points.

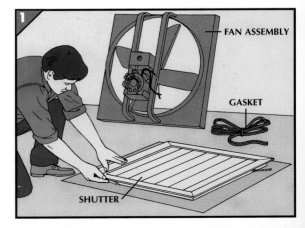

FAN ASSEMBLY

GASKET

SHUTTER

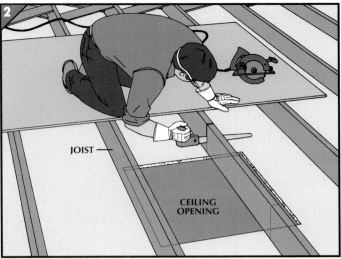

### Roof Trusses

*If your house has roof trusses (below) instead of the rafters and joists that are shown here—and if your fan does not fit between two trusses—do not cut them yourself. Consult a structural engineer before you proceed.*

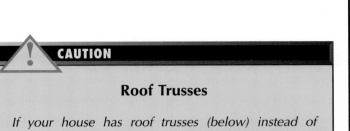

## 2. Cutting the opening.

◆ In the room below, align the template with the holes and mark the two other corners of the template on the ceiling. Drill small holes at these points.

◆ Join the four holes with straight lines to form a square and cut along the lines with a keyhole or saber saw. Skip over the ceiling joist when you come to it; after sawing, score the uncut segments of the outline with a utility knife.

◆ Break up the ceiling board under the joist with a hammer and tear it away with your hands.

◆ Working in the attic, remove a section of the exposed joist $1\frac{1}{2}$ inches outside the edges of the opening. Cut through most of the joist with a circular or saber saw. Then finish the cuts with a keyhole saw—reversing the handle to avoid damaging the ceiling below.

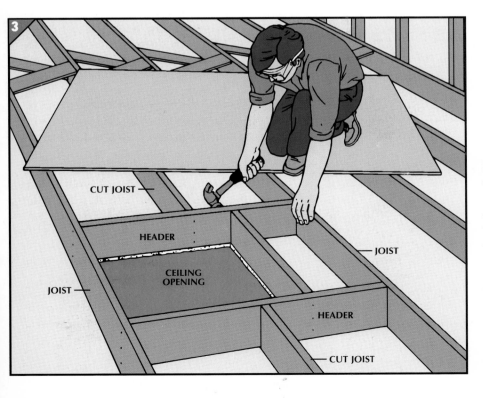

## 3. Making headers.

◆ Measure the distance between the uncut joists flanking the opening and cut two headers to this length, using wood the same size as the joists.

◆ Set the headers at the edges of the opening, and nail them to the sides of the uncut joists and the ends of the cut ones.

◆ To complete the frame, cut a third length of wood to fit between the headers, set it flush to the unframed edge of the opening, and nail it to the headers.

# INSTALLING THE FAN

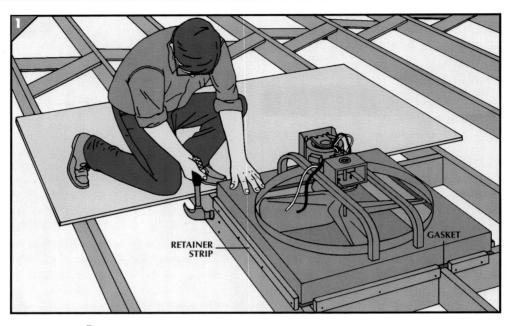

### 1. Positioning the fan assembly.
◆ Set the felt or rubber gasket on the wood frame (in some models this gasket is attached to the fan assembly at the factory) and then lower the fan onto the cushioned frame.
◆ Nail 1-by-3 retainer strips to the sides of the joists and headers around the fan assembly, with about an inch of each strip projecting above the frame.

### 2. Attaching the shutter.
◆ In the room below, adjust the spring at the side of the shutter to close the louvers gently when you hold them open and release them.
◆ Working from a stepladder, lift the vent into the ceiling opening and mark the positions of the screw holes in the shutter flange. Drill pilot holes at these points and screw the shutter through the ceiling to the frame above.
◆ In the attic, connect the black and white fan motor wires to a cable in an outlet box *(page 67, Step 3)* fastened to a nearby joist or header; from the box, run this cable to the basement service panel for a new 20-amp, 120-volt circuit connection *(pages 66-67, Steps 1-2)*.
◆ Install an on-off or a timer switch (the installation procedure is identical) on a convenient wall in the living quarters *(page 70, Step 5)*.

# Thrifty Cooling for Dry Climates

The ancient Egyptians, aware that evaporating water draws heat from the surrounding air, hung soaking-wet reed mats in doorways to cool entering breezes. A modern evaporative cooler operates on the same principle, using fiber pads instead of a mat and a blower for the breeze.

**How It Works:** Water enters a reservoir in the unit through a float-operated valve. When the reservoir is full, the float closes the valve to stop the flow. A pump then forces the water through tubes onto the pads. Most of the water evaporates into the air that is drawn through the pads by the blower; excess water drips from the pads back into the reservoir.

Evaporative coolers are satisfactory only in dry climates like the American Southwest. But where they are appropriate, coolers offer advantages over air conditioners. Because a cooler has no compressor, it is less expensive both to buy and to operate, using about a quarter as much electricity.

**Tips for Efficient Operation:** The air that is pumped into the house by the cooler must have an exit—usually opened windows—to keep the cool air flowing. Change the cooler's pads annually—or twice a year in areas where hard water clogs them with mineral deposits. Using artificially softened water does not help; minerals still remain in the water to choke the pads. Except in the case of sealed units, oil the blower motor and bearings regularly according to the manufacturer's instructions, and flush out the water reservoir as often as necessary to remove mineral deposits, sludge, and scum.

**TOOLS**

Carpenter's level
Screwdriver
Wrench

**MATERIALS**

Faucet adapter
Compression fittings
Copper tubing ($\frac{1}{4}$")

**Installing an evaporative cooler.**
◆ Attach the mounting brackets to the sill of a double-hung window.
◆ Attach the leveling-screw brackets to the bottom of the unit with their screws fully extended.
◆ With a helper, insert the outlet of the unit into the window. Hook the underside of the outlet to the window-sill brackets, and rest the plates at the heads of the leveling screws against the wall. Screw the plates to the wall.
◆ Close the window onto the unit, then level it by adjusting the leveling screws, and pull the outlet against the inside of the lower window sash. Close the gaps on the grille's sides with the adjustable panels that are supplied.
◆ Thread a faucet adapter onto an outdoor faucet. With compression fittings, attach $\frac{1}{4}$-inch tubing to the faucet adapter *(above)* and to the threaded inlet pipe at the back of the unit.

If no faucet is nearby, drill a hole near the unit through the side of the house into the basement. Run tubing from the unit to a cold-water pipe; connect the tubing to the pipe with a saddle valve and compression fittings.

# Economical Room Air Conditioners

Window- or wall-mounted air conditioners can cool more than one room, even an entire floor, provided the units have sufficient capacity and the cold air can be distributed from room to room. This approach is almost always less expensive than central air conditioning, even if a house requires several medium-size units to cool it effectively.

**Where to Put Them:** Plan air-flow patterns carefully before installing any air conditioners. The illustrations here show ways to circulate air with the use of fans, but if your house has a forced-air furnace, con-sider placing an air conditioner so that it blows cold air toward a return register of the heating system. Turn on the furnace blower to circulate cool air to every room in the house. This measure is even more effective if you increase the speed of the furnace blower during the cooling season.

Once you have chosen locations for the air conditioners, use the work sheet on page 124 and the manufacturer's recommendations to calculate the unit sizes you require. Add 10 percent to the area that each unit must cool to compensate for restrictions in air circulation.

**Installing a Window Unit:** The typical room air conditioner is a compact unit designed for quick window mounting with only a screwdriver. Alternatively, a window unit can be installed through an exterior wall *(pages 106-107)*.

Smaller units are usually sold with the internal machinery, called the chassis, premounted in a metal cabinet. Large air conditioners, however, are designed so that the chassis slides out of the cabinet to facilitate installation *(opposite)*. Specially shaped units are available for sliding windows *(page 106)* and for tall, narrow casement windows.

 **TOOLS**

Screwdrivers
Carpenter's level
Hacksaw
Electronic stud
 finder
Tape measure
Pry bar
Dry-wall saw
Handsaw
Electric drill
Circular saw
Hammer
Caulking gun

 **MATERIALS**

Adhesive-backed
 foam seal strips
 ($\frac{3}{8}$")
Wood screws
 ($\frac{3}{4}$" No. 8,
 1" No. 10)
Common nails (3")
Finishing nails (2")
Sponge foam
2 x 4s
2 x 6s
Plywood ($\frac{1}{2}$")
Shims
Silicone caulk
Molding

 **SAFETY TIPS**

*Wear a long-sleeved shirt, gloves, and a dust mask when you are removing insulation from a wall. Goggles protect your eyes when you are cutting into a wall and when you are hammering nails.*

# DIRECTING COLD AIR

### Cooling more than one room.
A portable fan set in a doorway pushes air from an air-conditioned room into an adjacent one. A 16-inch floor model works well; it circulates a large volume of air without creating unpleasant drafts or excessive noise.

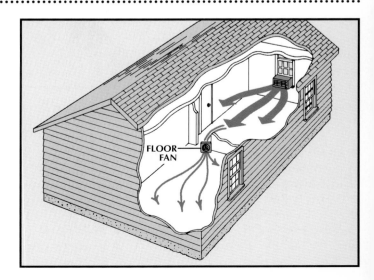

**Air-conditioning an entire floor.**
A large window unit at the end of a central hallway blows some cool air into the room at the opposite end of the hall, but fans are necessary to pull the airstream into the rooms on the right-hand side of the corridor. The room on the left-hand side of the hall has its own air conditioner because the door is frequently closed. A second air conditioner is also an option if cool air must be forced past more than two doorways or around more than two sharp corners. By closing doors of some rooms and turning off fans, you can channel cool air to where it is needed most.

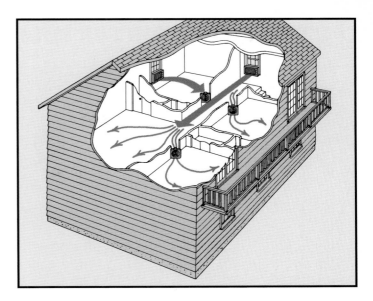

# A UNIT IN A DOUBLE-HUNG WINDOW

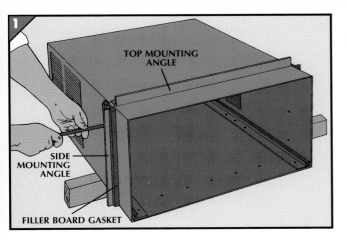

## 1. Preparing the cabinet.
◆ Remove the front grille and slide out the chassis. To do so, you may have to remove a bolt that secures the compressor to the cabinet.
◆ To prevent air leakage, apply a strip of $\frac{3}{8}$-inch adhesive-backed foam seal, provided with the unit or purchased at a hardware store, on the bottom of the cabinet where it will rest in the window frame. Prop the cabinet on a piece of scrap wood to avoid damage to the bottom seal as you work.
◆ Mounting angles are sometimes factory installed. If they have not been, press another strip of adhesive seal over the predrilled holes in the cabinet's top and attach the top angle.
◆ Fit the side angles into the filler board gaskets, which seal the angles to the cabinet, and screw on the angles.

## 2. Mounting the supports.
Units that extend more than a foot beyond the window stool require exterior support brackets. The horizontal arms of the brackets usually have a number of predrilled holes to allow for various window stool widths.
◆ Attach the horizontal legs to the underside of the cabinet through the factory-drilled holes, then bolt the vertical leg through the first hole beyond the stool and the angled leg as close to the outside end as possible.
◆ Position the cabinet on the stool so the lower ends of the legs rest against the house. As a helper supports the cabinet from the outside, fasten the cabinet and horizontal legs to the stool with $\frac{3}{4}$-inch No. 8 wood screws.
◆ Lay a carpenter's level on the cabinet above each bracket. Adjust the leveling screws until the bubble is half off its centerline, indicating a slight downward tilt.
◆ Fasten the cabinet to the stool with $\frac{3}{4}$-inch wood screws, then slide in the chassis and reaffix the grille.
◆ Add a strip of foam seal on the top angle, then lower the window sash; add filler boards at the cabinet sides, and stuff sponge foam between top and bottom sashes.

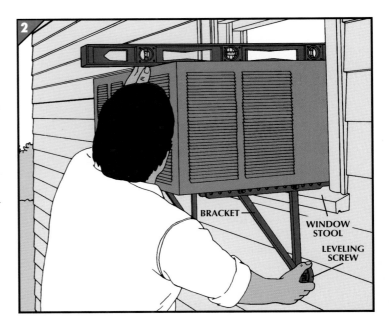

# SPECIAL INSTALLATION FOR SLIDING WINDOWS

RETRACTABLE FRAME

FILLER BOARD

## Filling the space.

A sliding window or crank-open casement window needs a special air conditioner that is slimmer and taller than a unit for a double-hung window. It contains a retractable frame that can be pulled upward out of the cabinet to the top of the window frame. When filled with thin filler board or translucent plastic, the frame seals off the area above the air conditioner and provides support at the top of the window frame. Where a frame is not provided, cut a piece of $\frac{1}{8}$-inch hardboard to fit the opening.

Fasten the unit to the window frame with the side mounting angles, which may require the removal of a casement window's cranking mechanism. If this is done, wedge a piece of wood between the cabinet and the open window to prevent the window from swinging against the air conditioner.

# A TECHNIQUE FOR PIERCING WALLS

## 1. Opening the wall.

◆ Mark the locations of studs on the wall where you will mount the unit; then mark the width of the air-conditioner cabinet on the interior wall.
◆ Remove the baseboard, and with a dry-wall saw cut the wallboard from floor to ceiling along the studs to either side of the air-conditioner location. Carefully pry off the wallboard so you can reuse it. Remove insulation from the wall cavity.
◆ From inside, mark the exterior sheathing for the cabinet opening, making it $\frac{1}{2}$ inch higher and wider than the cabinet dimensions.
◆ Drill four small holes at the corners. From the outside, use the holes as guides to saw through the siding and sheathing.
◆ If the wall is a bearing wall—the ceiling and floor joists run perpendicular to it—frame a temporary wall of 2-by-4s between floor and ceiling about 2 feet from the wall.
◆ Saw through the studs within the opening *(right)* and pry out the pieces.

## 2. Framing the opening.

◆ Trim two studs to fit between the top and sole plates, then cut two lengths of 2-by-4 as jack studs to reach from the sole plate to the top edge of the cabinet opening. Nail the pieces together in pairs and install the assemblies flush with the sides of the cabinet opening, toenailing them to the top and sole plates.

◆ Cut two lengths of 2-by-6 lumber to span the jack studs, then sandwich a piece of $\frac{1}{2}$-inch plywood between them to make a header. Toenail the header atop the jack studs.

◆ Measure a pair of cripple studs to fit between the header and top plate, and toenail them in place.

◆ Subtract 3 inches from the distance between the bottom of the opening and the sole plate, then cut two cripple studs to this length. Also cut two lengths of 2-by-4 to fit between the jack studs. Nail both of these boards to the ends of the cripple studs to form a double sill.

◆ Toenail the cripple studs to the sole plate and the double sill to the jack studs *(right)*.

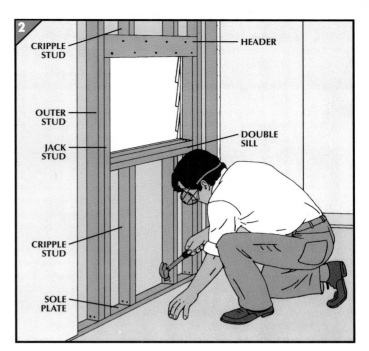

## 3. Installing the cabinet.

◆ Mount the exterior support brackets *(page 105)*, then slide the empty air-conditioner cabinet into the opening.

◆ Tap shims around the top, bottom, and sides of the cabinet to center it in the opening, then drill holes in the cabinet—two on the top and four on each side.

◆ Fasten the cabinet to the header and jack studs with 1-inch-long No. 10 wood screws.

## 4. Completing the installation.

◆ Caulk the cabinet edges, both inside and outside.

◆ Nail scrap 2-by-4s to the studs where the wallboard was cut away to provide a surface to reattach the material.

◆ Frame the cabinet inside the house with decorative molding, and replace the baseboard removed previously. You may also wish to outline the cabinet outside with molding to match that of nearby windows.

◆ With a helper, lift the air-conditioner chassis and slide it into the cabinet. Then attach the front grille.

With proper maintenance, the average window air conditioner will serve you well for many years. Once a year, clean the grille and the drain system. Gently vacuum both the evaporator and condenser coils (located at the front and back of the unit, respectively). At the same time, check the coil fins; vibration or extreme temperature change may have bent them out of shape. Most important, clean or replace your air filter at least once a month.

Except for changing the air filter, all of these maintenance chores require first removing the unit from the window. Since it is heavy, have someone help you.

⚠️ **CAUTION** *Avoid handling the coils or refrigerant lines; these carry high-pressure refrigerant and should be serviced only by a professional.*

 **TOOLS**

Nutdriver or socket wrench
Screwdriver
Multiheaded fin comb

 **SAFETY TIPS**

*Protect your hands with heavy work gloves when cleaning the coil fins.*

# CLEANING THE AIR FILTER

FRONT PANEL

### 1. Removing the front panel.
If you are just changing the air filter, unplug the air conditioner but leave it in the window. To take off the front panel, remove any screws, then pull the panel straight off. If the panel is secured by clips, grip its sides and snap it off *(left)*.

### 2. Removing the air filter.
On some models, the filter is located in front of the evaporator coils or on the back of the front panel; unfasten the retaining clips holding the filter in place and remove it *(right)*. On other models, the filter is mounted in front of the blower and below the evaporator coils; unfasten the filter from its clips and pull it out.

AIR FILTER

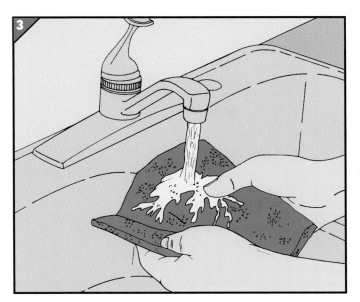

### 3. Washing the filter.

Each month during the cooling season, vacuum surface dirt from the filter, then wash it in a detergent-and-water solution, rinse it with fresh water, and wring it dry. If the filter is not washable or is torn, replace it with an identical filter.

### 4. Cleaning the front panel.

Use a moist cloth or stiff-bristled brush to wipe accumulated dust off both sides of the grille and the louvers of the front panel. Wash the panel in a detergent-and-water solution to remove greasy dirt, then rinse it in clear water and dry it. Reinstall the front panel and plug in the air conditioner.

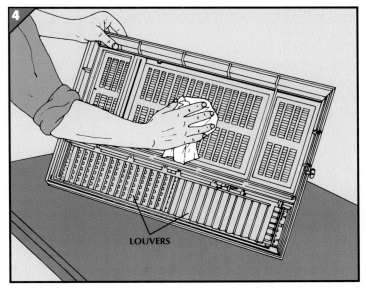

LOUVERS

# ACCESSING INTERNAL COMPONENTS

CHASSIS

### Freeing a large unit.

Most larger, heavier air conditioners, such as the one at left, have a slide-out chassis that fits into a cabinet bolted permanently to the window frame. Place a sturdy table next to the window, under the air conditioner, and work with a helper to remove the unit. Keeping your back straight and your knees bent to avoid muscle strain, slide the chassis out of the window onto the table.

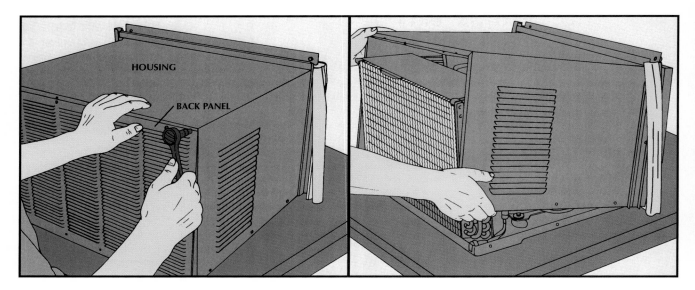

HOUSING

BACK PANEL

**Opening a smaller unit.**
Many small air conditioners have sliding panels at the sides, and the unit may be supported by a bracket outside the window. Have a helper support the unit while you carefully open the window.

Then collapse the side panels and ease the unit from the window. With a nutdriver, remove the sheet-metal screws securing the back panel to the chassis; if they are rusted, apply several drops of penetrating oil and use a sock-

et wrench to free them *(above, left)*. Next, remove the screws from the top and sides of the housing and, on some models, at the front. Grasp the bottom edges of the housing and pull it off *(above, right)*.

# CARING FOR THE COILS AND DRAIN SYSTEM

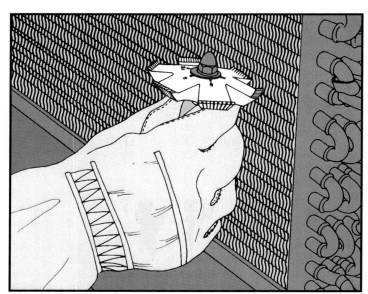

**Cleaning and straightening the coil fins.**
Vacuum the coil fins periodically with an upholstery-brush attachment. Use a multi-headed fin comb *(photograph)* to dislodge any debris stuck between the fins. Determine which head of the comb corresponds to the spacing of your unit's fins; the teeth on the head should fit easily between the fins. The comb can also be used to straighten bent coil fins. Gently insert the teeth into an undamaged section above the area to be straightened. Pull the fin comb down, sliding it through the damaged area *(above, right)*. Never straighten fins with a knife or screwdriver; they could damage the coils.

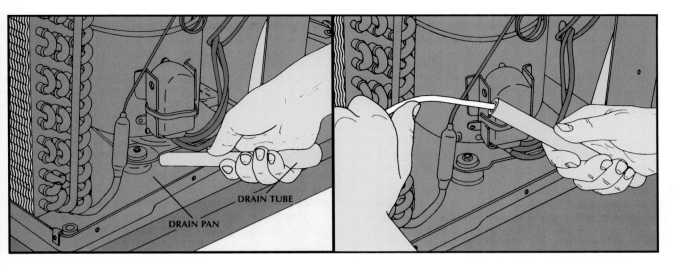

DRAIN TUBE

DRAIN PAN

## Unclogging the drain hole.
Locate the drain hole leading out of the chassis. If the hole is blocked, use a screwdriver or heavy wire to clear the hole *(below)*.

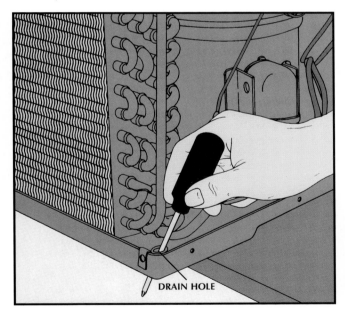

DRAIN HOLE

## Clearing the drain tube.
If your model has an evaporator drain pan and a condenser drain pan, locate the rubber or plastic drain tube connecting them. Pull the tube out from under the compressor base *(above, left)* and run a heavy wire through the tube to dislodge obstructions *(above, right)*. To prevent algae formation, flush the tube with a solution of 1 tablespoon of chlorine bleach in $\frac{1}{2}$ cup of water. If the air conditioner does not have a drain tube, use a cloth to wipe the drain channels molded into the drain pan.

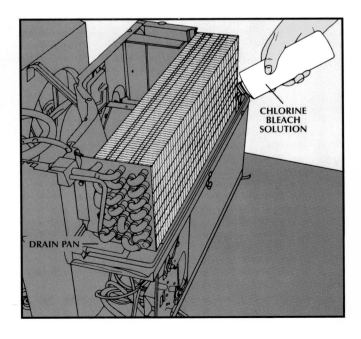

CHLORINE BLEACH SOLUTION

DRAIN PAN

## Cleaning the drain pans.
To prevent algae formation in the drain pans, place a bucket under the drain hole to catch overflow, and flush each pan with a solution of 1 cup of chlorine bleach and 1 cup of water. Rinse the pans with clear water.

Central air conditioners and heat pumps—essentially reversible air conditioners—operate similarly and share virtually the same servicing needs. Although repairs involving refrigerant can only be handled by a licensed professional, the procedures called for in the chart *(page 114)* and described in detail on the following pages pose no insurmountable difficulties to most homeowners.

**Annual Maintenance:** Central air conditioners and heat pumps consist of two basic parts: the con-denser unit, located outdoors, and the indoor evaporator coil. On a heat pump, the roles of these units reverse during the heating season. Check that the condenser unit is level to ensure proper operation. In the spring, clean the condenser coils and straighten any bent fins *(page 116)*. Lubricate the fan motor if necessary and inspect the fan for loose or deformed blades that can ruin the motor *(page 115)*.

Cold evaporator coils condense moisture out of the indoor air. This water drips into a drain pan beneath the coils. To prevent the growth of algae and bacteria in the pan and drain line, pour a 50/50 mixture of bleach and water into the pan and let it drain.

**A Unique Repair:** Heat pumps have a reversing valve that changes the direction of refrigerant flow to switch from cooling to heating and back. If your heat pump produces hot air in summer or the condenser coils do not deice in winter, check the reversing valve solenoid coil and replace it if it is defective *(page 123)*.

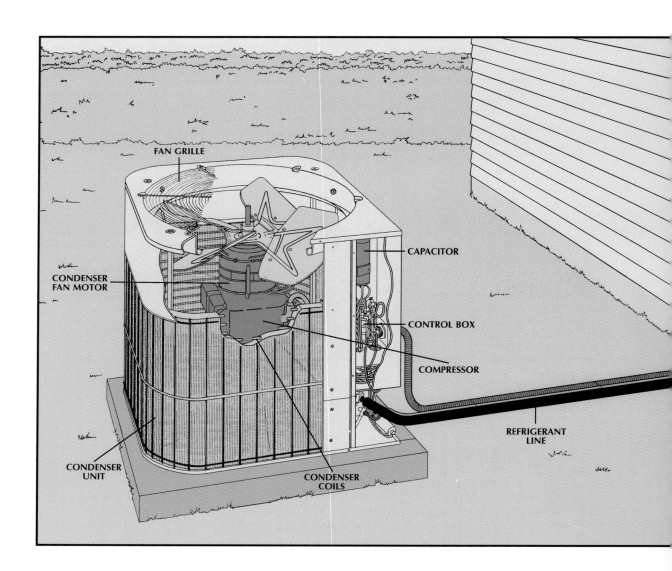

FAN GRILLE

CAPACITOR

CONDENSER
FAN MOTOR

CONTROL BOX

COMPRESSOR

REFRIGERANT
LINE

CONDENSER
UNIT

CONDENSER
COILS

**Shutting Down the System:**
Before beginning a repair, turn off power at the main service panel and at the switch near the condenser. Leave the unit off for at least 5 minutes before turning it back on to prevent excess refrigerant pressure from overloading the compressor.

 *Discharge all capacitors (page 117) before beginning the repairs on pages 118 to 123.*

**CAUTION**

 **TOOLS**

Screwdrivers
Wrenches
Fin comb
Jumper wires
Wire stripper

Resistor (20,000-ohm, 2-watt)
Multitester
Small toothbrush
Long-nose pliers
Emery board

**MATERIALS**

Light machine oil (nondetergent)

Contact cleaner
Crimp connectors

 **SAFETY TIPS**

*Protect your hands with heavy work gloves when removing panels from the condenser or when working on condenser fins. Wear rubber gloves when handling leaky capacitors (they may contain carcinogenic chemicals called PCBs), and insulated rubber gloves plus goggles when discharging a capacitor.*

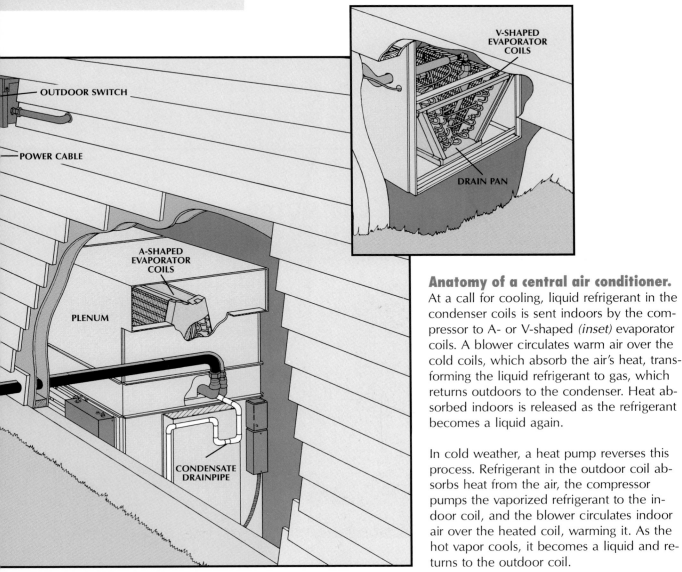

**OUTDOOR SWITCH**

**POWER CABLE**

**A-SHAPED EVAPORATOR COILS**

**PLENUM**

**CONDENSATE DRAINPIPE**

**V-SHAPED EVAPORATOR COILS**

**DRAIN PAN**

**Anatomy of a central air conditioner.**
At a call for cooling, liquid refrigerant in the condenser coils is sent indoors by the compressor to A- or V-shaped *(inset)* evaporator coils. A blower circulates warm air over the cold coils, which absorb the air's heat, transforming the liquid refrigerant to gas, which returns outdoors to the condenser. Heat absorbed indoors is released as the refrigerant becomes a liquid again.

In cold weather, a heat pump reverses this process. Refrigerant in the outdoor coil absorbs heat from the air, the compressor pumps the vaporized refrigerant to the indoor coil, and the blower circulates indoor air over the heated coil, warming it. As the hot vapor cools, it becomes a liquid and returns to the outdoor coil.

## Troubleshooting Guide

| PROBLEM | REMEDY |
|---|---|
| **Condenser unit does not turn on.** | Turn on outdoor switch; replace fuse or reset circuit breaker. <br> Lower thermostat setting. <br> Discharge capacitors *(page 117)*; test and replace if necessary *(page 118)*. <br> Test fan motor and replace if necessary *(page 122)*. |
| **Condenser unit does not turn off.** | Clean contactor *(pages 119-120)*; test and replace if necessary *(pages 120-121)*. |
| **Condenser unit noisy.** | Tighten fan-grille screws *(below)* and coil-guard screws *(page 116)*. <br> Inspect fan blades; adjust or replace if necessary *(page 115)*. <br> Lubricate fan motor *(page 115)*. <br> Test fan motor and replace if necessary *(page 122)*. |
| **Air conditioner does not cool.** | Lower thermostat setting. <br> Clean coils and remove any obstruction *(pages 116-117)*. <br> Test fan motor and replace if necessary *(page 122)*. |
| **Water leaking inside furnace.** | Clean drain pan. <br> Clean out trap in drainpipe. |
| **Heat pump produces hot air in summer or does not deice in winter.** | Test solenoid coil and replace if necessary *(page 123)*. |

# CHECKING THE CONDENSER FAN AND MOTOR

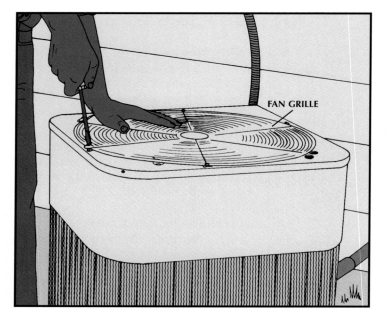

FAN GRILLE

**Removing the fan grille.**
◆ Turn off power to the condenser unit at the main service panel and outdoor-unit disconnect switch.
◆ To gain access to the fan, remove the screws holding the fan grille to the top panel *(left)*, and lift off the grille.

## Replacing fan blades.

◆ Wearing work gloves, spin the blades by hand. If they feel loose, or wobble as they turn, tighten the setscrew that secures them to the motor shaft.

◆ If the problem persists or if a blade is bent, loosen the setscrew and then lift the blade assembly off the motor shaft *(right)*.

◆ If necessary, oil the fan motor while the blades are removed *(below)*.

◆ Purchase replacement fan blades from an air-conditioning supplies dealer.

◆ Slide the blade assembly onto the motor shaft, aligning the setscrew with the flat side of the shaft.

◆ Hold the assembly about an inch above the motor, so that the blades clear the motor bracket, and tighten the setscrew.

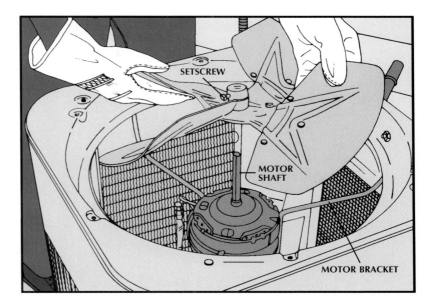

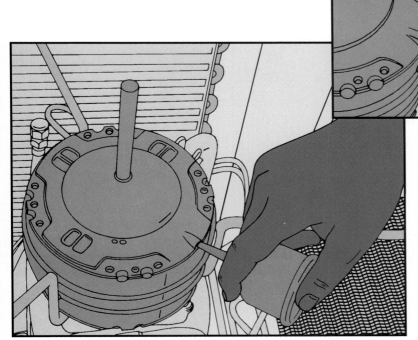

## Oiling the fan motor.

◆ Remove the fan grille *(left)* and blades *(above)*.

◆ Locate the oil ports, small holes in the motor housing. The ports may be sealed by plastic plugs; if so, pry out the plugs with a screwdriver *(inset)*.

◆ Inject 2 or 3 drops of nondetergent, light machine oil into each port *(left)*; do not overoil.

◆ Replace the plugs.

# FLUSHING THE CONDENSER COILS

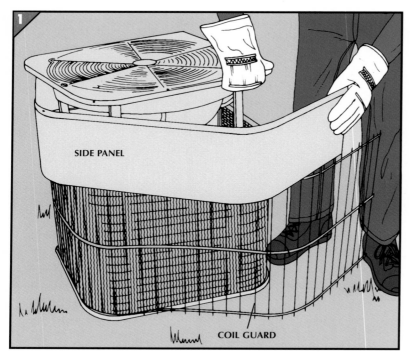

SIDE PANEL

COIL GUARD

## 1. Removing the coil guard.
◆ Turn off power to the condenser unit at the service panel and outdoor switch.
◆ Take out the screws connecting the side panel to the condenser unit top panel and frame. As on the model shown here, you may also have to remove the screws that secure the top of the unit to support rods inside.
◆ Work the side panel and coil guard loose and unwrap them from the condenser unit frame.

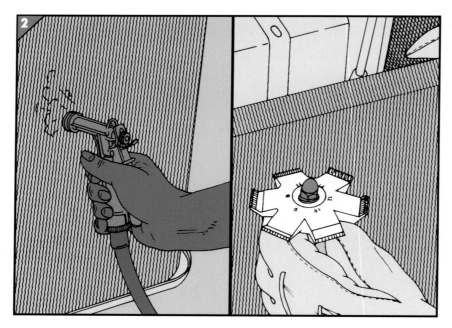

## 2. Refurbishing condenser fins.
◆ With a garden hose, spray water from inside the unit outward, then rinse the outside of the fins to dislodge dirt *(far left)*. Avoid spraying directly into the condenser fan motor. Clean greasy or exceptionally dirty coils chemically with a spray-on/rinse-off solvent made specifically for this purpose.
◆ Straighten any bent fins with a fin comb *(near left)*.

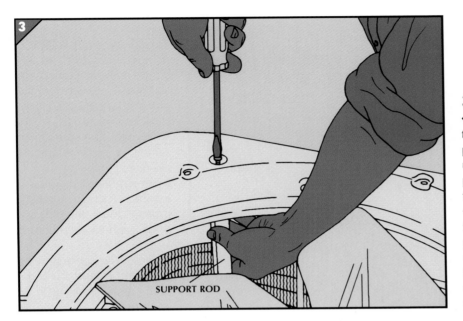

SUPPORT ROD

## 3. Reinstalling the panels.

◆ Screw one end of the side panel to the condenser frame. Wrap the panel and coil guard around the unit and, if necessary, under the top panel. Then fasten the side panel at the other end.

◆ If you removed screws from the top panel earlier, take off the fan grille *(page 114)*.

◆ Reach into the unit and hold each top-panel support rod in position while screwing the panel to it *(left)*.

◆ Reattach the grille.

# AN ELECTRICAL PRECAUTION

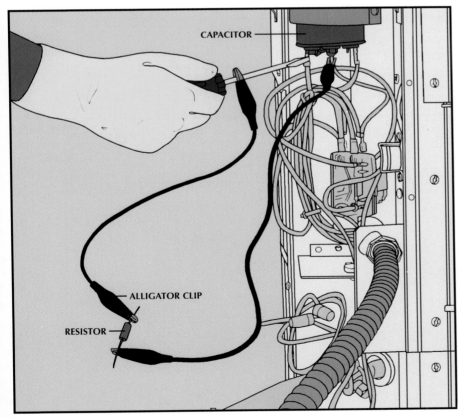

CAPACITOR

ALLIGATOR CLIP

RESISTOR

## Discharging a capacitor.

◆ To make a capacitor-discharging tool, buy two jumper wires fitted with insulated alligator clips and a 20,000-ohm, 2-watt resistor at an electronics-supply store. Clip one end of each jumper to the leads of the resistor, then clip one jumper to the blade of a screwdriver having a plastic handle.

◆ Turn off power to the condenser at the main service panel and outdoor switch. Turn off power to the indoor unit, as well.

◆ Remove the control box cover, located near the entry point of the power cable on the outdoor unit, and find the capacitor.

◆ Clip the free jumper to the common terminal of the capacitor.

◆ Hold the screwdriver handle in one hand, put your other hand behind you, then touch the blade to each capacitor terminal for 5 seconds *(left)*.

◆ Look for other capacitors in the unit and discharge them the same way.

# REPLACING A CAPACITOR

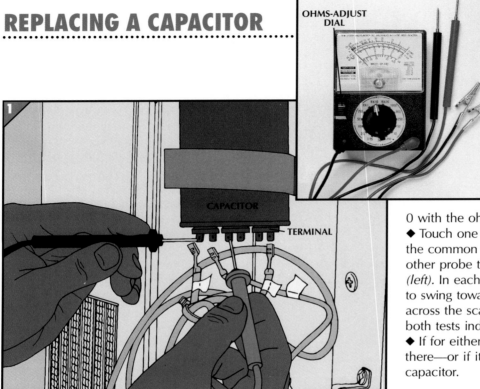

OHMS-ADJUST DIAL

CAPACITOR

TERMINAL

## 1. Testing a capacitor.

◆ Turn off power to the condenser at the main service panel and outdoor switch and power to the indoor unit. Discharge the capacitors *(page 117)*.

◆ To test a three-terminal capacitor, as shown, label and disconnect its wires.

◆ Set a multitester *(photograph)* to RX1K to measure resistance, touch the probes together, and set the needle to 0 with the ohms-adjust dial.

◆ Touch one probe to one of the connectors on the common terminal, marked C, and touch the other probe to both of the other terminals in turn *(left)*. In each case, watch for the multitester needle to swing toward 0 resistance, then slowly move across the scale toward infinity. This response in both tests indicates a serviceable capacitor.

◆ If for either test the needle swings to 0 and stays there—or if it does not move at all—replace the capacitor.

If the condenser has multiple, two-terminal capacitors, test them individually, touching the multitester probes to both terminals of each.

## 2. Connecting a new capacitor.

◆ Loosen the bracket *(right)* and slide out the capacitor.

◆ Purchase an identical replacement from a heating-and-cooling supplies dealer.

◆ Slide in the new capacitor and tighten the bracket.

◆ Reconnect the wires to the correct terminals and restore power.

⚠ **CAUTION** *Dispose of capacitors according to local hazardous-waste regulations.*

# MAINTAINING THE CONTACTOR

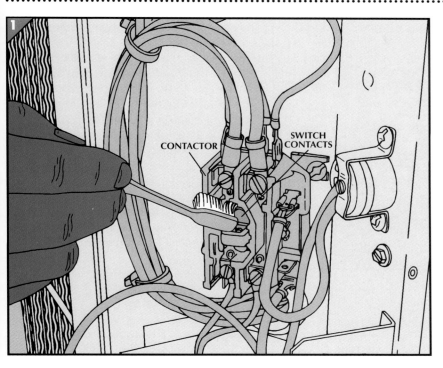

## 1. Cleaning the contacts.

◆ Turn off power to the condenser at the main service panel and outdoor switch, and shut off power to the indoor unit.

◆ Discharge the capacitor *(page 117)*.

◆ Locate the contactor and inspect the reed-style switch contacts. If they are stuck together or the spring is missing, replace the contactor *(page 121, Step 3)*.

◆ Clean the contacts with a small toothbrush moistened with contact cleaner *(left)*. Alternatively, spray contact cleaner onto the contacts with the extension tube provided with the cleaner.

◆ Reconnect loose wires, reassemble the unit, and turn on the power; if the contactor still doesn't work, test it *(pages 120-121, Steps 1 and 2)*.

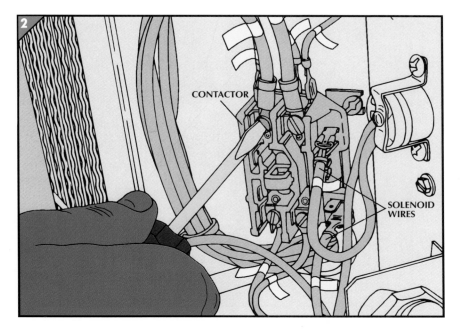

## 2. Disconnecting the wires.

◆ Turn off power to the condenser at the main service panel and outdoor switch. Turn off power to the indoor unit, as well.

◆ Remove the control box cover and discharge the capacitor *(page 117)*.

◆ Label the contactor wires, including the wires for the solenoid coil behind the contactor, and disconnect them *(left)*; pull wires from spade connectors with long-nose pliers. For easier access, you can unscrew the contactor bracket and remove the contactor from the control box.

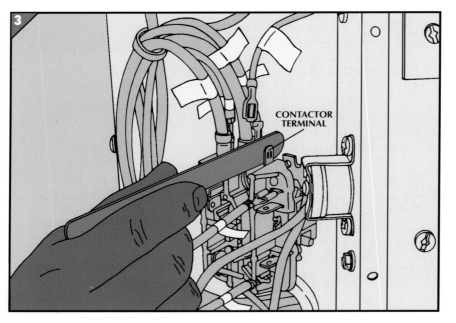

**CONTACTOR TERMINAL**

### 3. Cleaning the terminals.
◆ Inspect the screw terminals and spade connectors for dirt or corrosion. Brighten the metal by rubbing with an emery board *(left)*.
◆ Examine the connectors at the ends of the wires you removed in Step 2. Cut off any loose-fitting or damaged connector, strip away $\frac{3}{8}$ inch of insulation, and crimp a new, so-called solderless connector to the bare wire with the appropriate notch of a wire stripper tool. You can purchase UL-approved connectors at an electrical-parts supplier to fit the wires and terminals in your condenser.
◆ Reconnect each wire in turn, crimping the connector onto the terminal.

# A NEW CONTACTOR

### 1. Contacts-open test.
◆ Turn off power to the indoor and outdoor units, discharge the capacitors *(page 117)*, and set a multitester to RX1K.
◆ Disconnect the wires from one upper screw terminal.
◆ Clip one multitester lead to one of the upper terminals and touch the other lead's probe to the adjacent screw terminal *(right)*. Check that the multitester needle points to infinity.
◆ Disconnect the wires from one of the lower terminals and test the lower pair the same way.
◆ If the needle points to infinity in both tests, proceed to the next step. Otherwise, replace the contactor *(page 121, Step 3)*.

## 2. Contacts-closed test.

◆ Gently hold the upper switch contacts closed *(right),* and repeat the tests in the preceding step. Check that the multitester needle points to 0.

◆ Test the lower pair of switch contacts the same way.

◆ Replace the contactor *(below)* if the multitester needle does not point to 0 in either test.

◆ If the contacts are okay, test the solenoid coil behind the contactor by first disconnecting the upper and lower wires from the solenoid coil terminals. Set a multitester to RX1K and touch the probes to the terminals. If there is no continuity, replace the contactor.

◆ Reattach the control box cover and restore power to the unit.

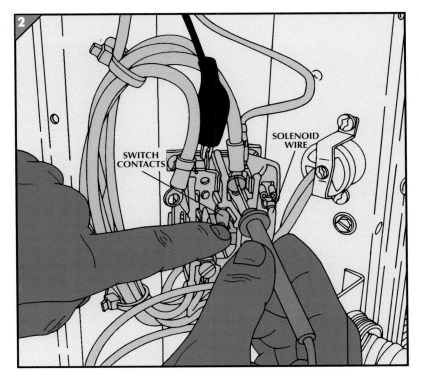

SWITCH
CONTACTS

SOLENOID
WIRE

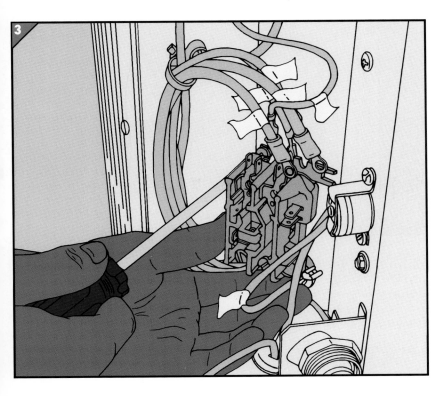

## 3. Replacing the contactor.

◆ With power to the indoor and outdoor units turned off and the capacitor discharged, label and disconnect all contactor wires.

◆ Remove the mounting screws holding the contactor bracket to the control box wall *(left)* and take out the contactor.

◆ Purchase an identical replacement contactor from a heating-and-cooling supplies dealer.

◆ Screw the new contactor in place and reconnect the wires.

◆ Reassemble the unit and turn on the power.

# TESTING AND REPLACING THE FAN MOTOR

## 1. Testing for continuity.

◆ Turn off power to the condenser unit at the main service panel and outdoor switch. Turn off power to the indoor unit, as well.

◆ Take off the fan grille *(page 114)*, control box cover, and the fan blades *(page 115)*.

◆ Discharge the capacitor *(page 117)*.

◆ Trace the wires leading from the fan motor to the contactor. If the motor has four leads, two will go directly to the contactor. If there are three leads, one will go to the contactor and another will go first to the common lead on the capacitor.

◆ Disconnect one of the fan wires from the contactor. Label and disconnect other wires in the way.

◆ Set a multitester to RX1K and touch one probe to each fan wire *(right)*. Check that the needle points to 0.

◆ Touch one probe to the motor housing and the other to each fan wire. Check that the needle points to infinity.

◆ If the fan motor fails any test, replace it *(below)*.

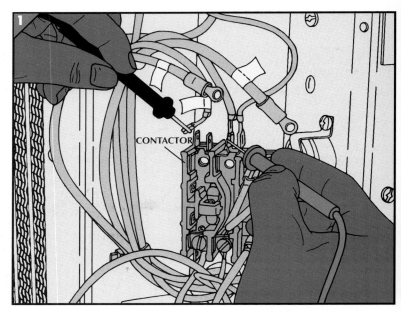

CONTACTOR

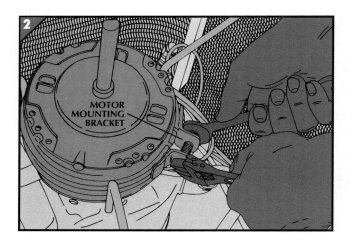

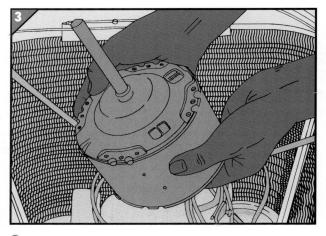

## 2. Loosening the bracket.

In the model above, the fan motor is suspended in a bracket at the center of the condenser. If the motor is hot, wait for it to cool.

◆ Before unbolting the bracket, pull the motor wires through from the control box; you may have to release some wire-retaining bands first.

◆ Loosen, but do not remove, the nut on the mounting bracket bolt with two wrenches *(above)*.

## 3. Removing the motor.

◆ If there is a retaining screw in the mounting bracket that is set against the motor to hold it steady, loosen it with a screwdriver.

◆ Spread the bracket apart, and push up on the motor from below. Then lift it from the bracket, holding it firmly by the housing, not the shaft *(above)*.

◆ Take the motor with you to purchase a replacement from a heating-and-cooling supplies dealer.

◆ Install the new motor, reversing the steps taken to remove it.

◆ Reattach the fan blades *(page 115)*, re-assemble the unit, and turn on the power.

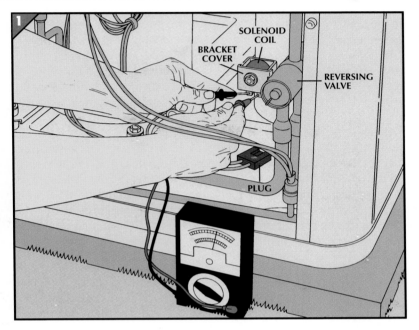

## 1. Testing the solenoid coil.

◆ Remove the rear access panel and find the solenoid coil attached to the reversing valve, a cylindrical device with one refrigerant pipe extending from the top and three from the bottom.

◆ With your fingers or long-nose pliers, disconnect the plug to the solenoid coil.

◆ Set a multitester to RX1K. Reaching under the solenoid bracket cover, touch the probes to the solenoid coil terminals. Check that the meter registers a value other than 0 or infinity.

◆ Replace the coil if it fails this test *(below)*.

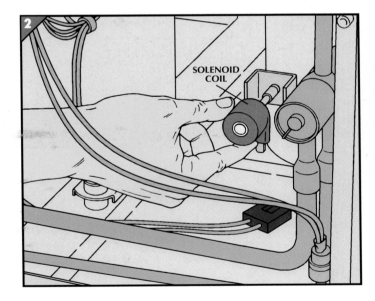

## 2. Replacing the solenoid coil.

◆ With a wrench, remove the nut that holds the solenoid bracket cover and coil in place.

◆ Pull the metal bracket cover away from the coil assembly, and slide the old coil off the shaft and replace it with a new one.

◆ Screw the access panel back into place.

# Planning for Comfort and Efficiency

When replacing a furnace, air conditioner, or heat pump, a new unit the same size as the old one can be a poor choice; it fails to take into account the aging of the house or improvements that may require different heating or cooling capacity.

A unit that is too small simply won't do the job, and one that is too large wastes money. In the case of an air conditioner, which dehumidifies the air in addition to cooling it, overcapacity can result in an uncomfortably clammy atmosphere: The unit cools the house before it has time to dry the air.

A Good Contractor: Buying the right size heating and cooling system is a complicated issue that is best left largely to a contractor. Find one who will agree to make a thorough assessment of your house and use the results to make a recommendation.

You can do some of the work yourself by filling in the house survey below. Using the map and charts at right, you can also assess insulation requirements, a factor that is as important to sizing a heating and cooling plant as it is to saving money on heating and cooling bills.

Other Factors: Even with this information from you, the contractor has many aspects of the house yet to investigate. Among them, for example, are the number of heating and cooling zones (or thermostats) in the house, and the condition of weather stripping and ducts. (Leaky ducts can rob a system of up to 30 percent of its capacity.) These factors and others, all influenced by the severity of your climate, yield a figure in BTUH required to heat or cool your house.

A Matter of Efficiency: No furnace, air conditioner, or heat pump is 100 percent efficient in using this energy. Yet the unit that wastes the least is not always the best choice. Often more expensive than a less efficient model, it may not pay for itself in fuel savings before you plan to sell the house. For this reason and others, your contractor may recommend something other than a unit with the highest efficiency rating.

## Hiring a contractor.

Filling out this inventory will give a heating and cooling contractor a head start in evaluating your house for a new heating and cooling system. While most of the entries are unambiguous, a couple of them benefit from explanation. Under window types, for example, count storm windows as single pane, not double. If a ceiling or floor has no more than one-tenth the insulation recommended in the charts on the next page, count the area as uninsulated.

## HOUSE INVENTORY

### Window Area by Side of House
North: _____ sq. ft.
South: _____ sq. ft.
East: _____ sq. ft.
West: _____ sq. ft.

### Window Types
Single pane: _____ sq. ft.
Double pane or glass block: _____ sq. ft.

### Walls
Length of all exterior walls (southern exposure): _____ ft.
Length of all exterior walls (all other exposures): _____ ft.
Length of interior walls separating a garage or
    other unheated or uncooled space from the house: _____ ft.

### Area of Roof or Ceiling
Roof, uninsulated: _____ sq. ft.
Roof with at least 1" of insulation: _____ sq. ft.
Ceiling with occupied space above: _____ sq. ft.
Ceiling below an insulated attic: _____ sq. ft.
Ceiling below an uninsulated attic: _____ sq. ft.

### Floor (Disregard if house is on slab)
Insulated: _____ sq. ft.
Uninsulated: _____ sq. ft.

# HOW MUCH INSULATION IS ENOUGH?

**Insulation zones.**
This map and accompanying chart reflect the U.S. Department of Energy's insulation recommendations, expressed as R-values, for each of three insulation zones in the United States. Irregular zone contours result in part from the generally lower temperatures found in mountainous regions. All of Canada lies in Zone 3, except for the west coast, which falls in Zone 2. If you live on the border between two zones, choose the zone with the higher number.

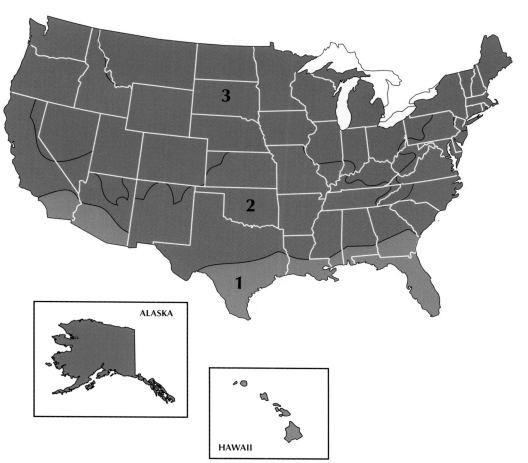

## RECOMMENDED R-VALUES

| Zone | Ceiling or roof | Exterior wall | Floor |
|------|-----------------|---------------|-------|
| 1 | R-30 | R-11 | 0 |
| 2 | R-38 | R-11 | R-19 |
| 3 | R-49 | R-11 | R-19 |

## R-VALUES BY THICKNESS OF INSULATION

| R-Value | Mineral Fiber Batts or Blankets | Fiberglass Batts or Blankets | Cellulose Fiber Loose or Blown Fill |
|---------|---------------------------------|------------------------------|-------------------------------------|
| R-11 | $3\frac{1}{2}$" | $4\frac{3}{4}$" | $3\frac{3}{4}$" |
| R-19 | 6" | 8" | $6\frac{1}{2}$" |
| R-30 | $9\frac{1}{4}$" | $12\frac{1}{2}$" | $10\frac{1}{2}$" |
| R-38 | $11\frac{1}{2}$" | 16" | 13" |
| R-49 | $15\frac{1}{4}$" | $20\frac{1}{2}$" | 17" |

**Assessing your insulation.**
Any house benefits from insulation installed in any partition—ceiling, roof, floor, or wall—that marks a boundary of the heated and air-conditioned parts of a house. To find out how much insulation you have, check unfinished attics, basements, or crawlspaces. Insulation batts and blankets usually have the R-value printed on the backing paper or foil. If not—or if you have loose, blown-in insulation—measure the thickness of the insulation and consult the chart to find the approximate R-value.

R-value can be increased in attics by laying new batts or blankets perpendicular to the existing material until the combined R-values equal the recommended number. In walls or floors, the cost of ripping out dry wall or plaster must be factored against the benefit of lower energy costs over the period of time you expect to live in the house.

# INDEX

10; balancing, 10-11; boiler, 48; expansion tank, 51; pressure, 48; tapping for fan-coil heater, 80, 81; water-flow patterns, 75. *See also* Expansion tank; Hot-water system pump

Hot-water system, maintenance: 48; adding shutoff valve, 51; draining and refilling, 48; expansion tank, 51; pressure, 48; quieting convector, 11; straightening convector fins, 11; troubleshooting chart, 49. *See also* Hot-water system pump

Hot-water system pump: low-maintenance, 48, 50; lubricating, 48; replacing coupler, 50; replacing motor, 49; replacing seal, 50

Humidifier, bypass: 38; maintenance, 38-39

Humidifier, fan-powered: 38; maintenance, 40

Humidifier, installing fan-powered: 38; estimating capacity, 41; humidistat, 42; materials and tools for, 38; mounting, 41; water connections, 42, 43; wiring, 43

Humidity control: with energy-recovery ventilator, 47

### I

Insulation: ratings, 125; types, 125; zones, 125

### J

Jump (ductwork): 45

### L

Ladder hooks: 89
Ladders: using on roof, 89
Liner: chimney, 26; combus-

tion chamber, 28

### O

Oil burner: anatomy, 29; functioning, 28; nozzle specifications, 28, 32

Oil burner, maintenance: 28; chimney and combustion chamber, 28; cleaning fan, 31; cleaning flame sensor, 31; cleaning ignition system, 32; cleaning nozzle, 32-33; cleaning oil lines, 33; efficiency check, 28; firing assembly maintenance, 32-33; lubricating motor, 31; materials and tools for, 29; preparation, 28; priming pump, 33; relining combustion chamber, 28, 34-35; removing water from tank, 28; sealing furnace leaks, 34; troubleshooting chart, 29; unclogging filter and strainer, 30

### P

PCBs: and capacitors, 113
Pilot light, furnace: 18; adjusting, 23-24; lighting, 23
Pipe: adding valve to copper, 51; adding valve to steel, 51; for hot-water heating, 75; joining copper and steel, 75, 78-79. *See also* Ducts
Plenum: 8
Pool heater, solar: 93

### R

Rain cap, chimney: 27
Refrigerant: repairs involving, 112
Registers, adding: cutting holes, 55-56; making extension

boxes, 57; making templates, 54-55; mapping ducts, 54; materials and tools for, 54; positioning in new line, 58, 62; register types, 56

Reversing valve: 112, 123

Roofs: cutting for fan, 97; truss-type, 101; working on, 88, 89

### S

Safety: with asbestos, 48; with cooling lines or fins, 108; drilling, 88; with ducts, 58; with electrical wiring, 36; with gas furnace, 18, 19; hammering, chiseling, and sawing, 66; with insulation, 100; with linoleum knives, 96; with PCBs, 113; with pipe, 75; with pipe cement, 93; on roofs, 88; with sheet metal, 8; soldering, 88; working on air conditioners, 112, 113; working on ceiling openings, 58

Sail switch: 46

Screwdriver, flexible. *See* Spinner

Service panel: wiring at, 66

Solar heat. *See* Trombe wall

Solar pool heater. *See* Pool heater, solar

Solar water heater. *See* Water heater, solar

Soldering: on roof, 90; safety, 88; and washers, 91

Spinner: 22

Stud finder, electronic: 66

### T

Thermostats: adjusting, 13; anatomy, 12; bypassing faulty, 13; for heat pump, 12; insulating, 14; line-voltage,

12; programmable, 12

Thermostats, installing: line-voltage, 14; materials and tools for, 12; programmable, 15

Thermostats, repairing: 12-13; adjusting anticipator, 13; blocking drafts, 14; cleaning, 13; materials and tools for, 12; releasing jams, 12; tightening wires, 12

Trombe wall: anatomy, 83; functioning of, 82; measurements, 82, 83; requirements for, 82

Trombe wall, installing: 82; flashing, 86; frame, 84-85; materials and tools for, 82; panels, 86; planning, 83; vents, 84, 87

Troubleshooting charts: for central air conditioners, 114; for gas furnaces, 19; for heat pump, 114; for hot-water system, 49; for oil burner, 29

Truss roofs: 101

### V

Vent, gas heater: 74
Venturi T connector: 75, 77

### W

Walls: cutting masonry, 84; cutting through siding, 106; suited for Trombe wall, 82

Water heater: tapping for heat, 80, 81

Water heater, solar: anatomy, 88; siting, 88.

Water heater, solar, installing: connecting to tanks, 91; locating, 89; materials and tools for, 88; starting system, 92; wiring, 92; working on roof, 89-90

Time-Life Books is a division of Time Life Inc.

PRESIDENT and CEO: John M. Fahey Jr.

## TIME-LIFE BOOKS

MANAGING EDITOR: Roberta Conlan

*Director of Design:* Michael Hentges
*Director of Editorial Operations:*
 Ellen Robling
*Director of Photography and Research:*
 John Conrad Weiser
*Senior Editors:* Russell B. Adams Jr.,
 Dale M. Brown, Janet Cave, Lee Hassig,
 Robert Somerville, Henry Woodhead
*Special Projects Editor:* Rita
 Thievon Mullin
*Director of Technology:* Eileen Bradley
*Library:* Louise D. Forstall

PRESIDENT: John D. Hall

*Vice President, Director of Marketing:*
 Nancy K. Jones
*Vice President, Director of New Product
 Development:* Neil Kagan
*Associate Director, New Product
 Development:* Elizabeth D. Ward
*Marketing Director, New Product
 Development:* Wendy A. Foster
*Vice President, Book Production:*
 Marjann Caldwell
*Production Manager:* Marlene Zack
*Quality Assurance Manager:* James King

## HOME REPAIR AND IMPROVEMENT

SERIES EDITOR: Lee Hassig
*Administrative Editor:* Barbara Levitt

Editorial Staff for *Heating and Cooling*
*Art Directors:* Barbara M. Sheppard
 (principal), Mary Gasperetti
*Picture Editor:* Catherine Chase Tyson
*Text Editor:* James Michael Lynch
*Associate Editors/Research-Writing:*
 Dan Kulpinski, Terrell Smith
*Technical Art Assistant:* Angela Miner
*Senior Copyeditor:* Juli Duncan
*Copyeditor:* Judith Klein
*Picture Coordinator:* Paige Henke
*Editorial Assistant:* Amy S. Crutchfield

*Special Contributors:* John Drummond
 (illustration); Jennifer Gearhart, Craig
 Hower, Marvin Shultz, Eileen Wentland
 (digital illustration); George Constable,
 Brian McGinn, Peter Pocock, Eric
 Weissman (text); Mel Ingber (index).

*Correspondents:* Christine Hinze (London),
 Christina Lieberman (New York), Maria
 Vincenza Aloisi (Paris).

## PICTURE CREDITS

*Cover:* Photograph, Renée Comet; Art,
 Patrick Wilson/Totally Incorporated.

*Illustrators:* George Bell, Roger Essley, Nick
 Fasciano, Gerry Gallagher, Great, Inc.,
 Dale Gustafson, Fred Holz, Mitchell
 Kuff, John Massey, Peter McGinn, John
 Sagan, Ray Skibinski, Tipy Taylor/Totally
 Incorporated, Vantage Art, Inc., Edward
 Vebell, Whitman Studio, Inc.

*Photographers:* **End papers:** Renée Comet.
 **14, 15, 21, 22:** Renée Comet. **27:**
 Renée Comet, prop courtesy Z-Flex
 U.S., Inc. **46:** Renée Comet. **50:** ITT
 Bell & Gossett. **56, 61, 66, 76, 84:**
 Renée Comet. **110:** Renée Comet, prop
 courtesy Robinair. **118:** Renée Comet.

## ACKNOWLEDGMENTS

Aireco Supply, Inc., Alexandria, Va.; John
Bittner, Chimney Safety Institute of Ameri-
ca, Gaithersburg, Md.; Tim Dougherty,
Dwyer Instruments, Inc., Michigan City,
Ind.; Ron Facchina, Automatic Equipment
Sales of Washington, Inc., Alexandria, Va.;
Aaron Gardner, Z-Flex U.S., Inc., Bedford,
N.H.; Jimmy Graham, Washington Gas,
Springfield, Va.; Tracy Haak, Association
of Home Appliance Manufacturers, Chica-
go; Bud Healy, North American Heating,
Refrigeration, and Airconditioning Whole-
salers Association, Columbus, Ohio; Home
Ventilating Institute, Arlington Heights, Ill.;
Honeywell Inc., Minneapolis; Richard
Kline, Washington Gas, Springfield, Va.;
Michael Lamb, Energy Efficiency and
Renewable Energy Clearinghouse, Merri-
field, Va.; John Lonigan, Aireco Supply,
Inc., Arlington, Va.; Ross McDaniel, J. & H.
Aitcheson, Inc., Alexandria, Va.; Glenn
Pottberg, Burnham Corporation, Lancaster,
Pa.; Hank Rutkowski, Air Conditioning
Contractors of America, Washington, D.C.;
Joel Sack, Automatic Equipment Sales of
Northern Virginia, Herndon, Va.; Larry
Sheffield, American Metal Products Com-
pany, Olive Branch, Miss.; Dwight Shuler,
Owens-Corning Fiberglas, Toledo; Denny
Speas, NIBCO, Inc., Elkhart, Ind.; Joe Teets,
Fairfax, Va.; Scott Thompson, Flex-L
International, Inc., Worthington, Ohio;
Bill Wiley, Automatic Equipment Sales of
Washington, Inc., Alexandria, Va.; Williams
Furnace Company, Colton, Calif.

First printing. Printed in U.S.A.
Published simultaneously in Canada.
School and library distribution by Time-Life
Education, P.O. Box 85026, Richmond,
Virginia 23285-5026.

TIME-LIFE is a trademark of Time Warner
Inc. U.S.A.

**Library of Congress
Cataloging-in-Publication Data**
Heating and cooling / by the editors of
 Time-Life Books.
p.  cm. — (Home repair and improve-
 ment)
Includes index.
ISBN 0-7835-3897-9
1. Dwellings—Heating and ventilation—
 Amateurs' manuals. 2. Dwellings—Air
 conditioning—Amateurs' manuals.
I. Time-Life Books. II. Series.
TH7224.H43 1995
697—dc20                    95-31095